Human-Centred Systems in the Global Economy

ARTIFICIAL INTELLIGENCE AND SOCIETY

Series Editor: KARAMJIT S. GILL

Knowledge, Skill and Artificial Intelligence
Bo Göranzon and Ingela Josefson (Eds.)

Artificial Intelligence, Culture and Language:
On Education and Work
Bo Göranzon and Magnus Florin (Eds.)

Designing Human-centred Technology: A Cross-disciplinary
Project in Computer-aided Manufacturing
H.H. Rosenbrock (Ed.)

The Shape of Future Technology:
The Anthropocentric Alternative
Peter Brödner

Crossing the Border: The Social and Engineering Design of
Computer Integrated Manufacturing Systems
J. Martin Corbett, Lauge Baungaard Rasmussen and
Felix Rauner

Artificial Intelligence and Human Institutions
Richard Ennals

Dialogue and Technology: Art and Knowledge
Bo Göranzon and Magnus Florin (Eds.)

Understanding the Artificial:
On the Future Shape of Artificial Intelligence
Massimo Negrotti (Ed.)

Computer-aided Manufacturing and Women's Employment:
The Clothing Industry in Four EC Countries
Swasti Mitter (Ed.)

Connectionism in Context
Andy Clark and Rudi Lutz (Eds.)

Yuji Masuda (Ed.)

Human-Centred Systems in the Global Economy

Proceedings from the International Workshop on Industrial Cultures and Human-Centred Systems held by Tokyo Keizai University in Tokyo, 1990

With Introduction by Mike Cooley

With 20 Figures

Springer-Verlag
London Berlin Heidelberg New York
Paris Tokyo Hong Kong
Barcelona Budapest

Yuji Masuda, BA, BE, ME
Faculty of Business Administration
Tokyo Keizai University
1-7 Minami-cho
Kokubunji-Shi
Tokyo 185, Japan

British Library Cataloguing in Publication Data
Human-Centred Systems in the Global Economy: Proceedings from the International Workshop on Industrial Cultures and Human-Centred Systems held by Tokyo Keizai University in Tokyo, 1990. - (Artificial Intelligence and Society Series)
I. Masuda, Yuji II. Series
006.3

Library of Congress Cataloging-in-Publication Data
International Workshop on Industrial Cultures and Human Centred Systems (1990 : Tokyo, Japan)
Human centred systems in the global economy : proceedings from the International Workshop on Industrial Cultures and Human Centred Systems, held by Tokyo Keizai University in Tokyo, May 1990 / Yuji Masuda (ed.) ; with introduction by Mike Cooley.
p. cm. -- (Artificial intelligence and society)
Includes bibliographical references.

ISBN-13: 978-3-540-19745-4 e-ISBN-13: 978-1-4471-1967-8
DOI: 10.1007/978-1-4471-1967-8

1. Industrial sociology--Congresses. 2. Technology--Social aspects--Congresses. I. Masuda, Yuji, 1938- . II. Tokyo Keizai Daigaku. III. Title. IV. Series.
HD6952.I58 1990 92-10114
306.3'6--dc20 CIP

34/3830-543210 Printed on acid-free paper

Preface

This book originates from an international workshop held in Tokyo in May 1990 on Industrial Cultures and Human-Centred Systems. The workshop brought together researchers and practitioners from Japan, Europe, South East Asia, Eastern Europe, Soviet Union, and North America in the areas of industrial and production cultures, human-centred systems and anthropocentric technologies, organisational innovation, technological change and working life issues. The workshop also discussed contextual issues of industrial culture, society and technology, technology and the citizen in the modern society, technology and knowledge transfer.

The workshop was organised as part of the centenary celebrations of the Tokyo Keizai University, and was the first workshop on human-centred systems in Japan. The interest in industrial cultures, globalisation, and technology shaping is increasing not only in Europe but also in Japan and North America. The present volume is a contribution to the widening of the European debate on human-centred manufacturing systems to broader issues of industrial cultures, economic competitiveness and globalisation of knowledge. The international debate on these issues is covered in the international journal, *AI & Society*, and its sister book series on *Artificial Intelligence and Society*. The interest in societal and cultural issues in the volume is also a reflection of the current European debates on human-centredness and sustainable systems design, social and cultural shaping of technology, information technology and social citizenship.

The contributions to the volume vary in their styles of articulation and presentation, and have been largely left in their original form so as to reflect the cultural diversity and research traditions of the contributors. I very warmly appreciate the support and advice given by Karamjit S. Gill and Mike Cooley in bringing out this volume. The SEAKE Centre researchers Jim Thorpe, John Gøtze, David Smith assisted by Harinder Gill have supported me in coordinating the editing and the production of the manuscript. I am indebted to Tokyo Keizai University for sponsoring the workshop and this book.

Tokyo *Yuji Masuda*
November 1991

Introduction

Mike Cooley

The term *human-centred* was first used in 1976 when I described 'telechiric devices' which respond to human skill and ingenuity and did not objectivise or marginalise human capacities. They were, therefore, in the Heidegger sense, tools rather than machines. The term arose in the rich and diverse context of the Lucas Workers' Plan for socially useful, sustainable technology[1].

The Plan caused quite a stir at that time and was met with incredulity by those who should have known better.

In order to demonstrate in practice what might be possible, projects and prototypes were developed to demonstrate different aspects of the Plan. Many of the products proposed at that time are being manufactured worldwide as the growing consciousness of environmental issues requires a dramatic realignment of what we perceive economic feasibility to be.

Demonstrating the wider concepts of human-centredness proved, however, to be much more difficult. It was not until 1983 that a multidisciplinary team of engineers, social scientists, philosophers and academics, linked with industrialists in Denmark, Germany and Britain approached the EEC for funding to design and demonstrate the world's first Human-Centred Computer Integrated Manufacturing System (ESPRIT Project 1217). This inevitably meant that the project had to be highly focussed on production systems so that the attributes of such a system could be clearly demonstrated. This was done successfully but since most people heard of human-centred systems for the first time in a CIM context, they came to think of it as applying solely to manufacturing concerns. This was an unfortunate but understandable narrowing of the scope of the concept.

The great service that Tokyo Keizai University and Professor Yuji Masuda have done with this present book is to restore human-centred systems to their rightful, more universal context.

The book is also timely. Spectacular failures of machine centred systems, particularly in the United States, have brought into sharp focus the vulnerability and lack of robustness in those systems where the human capacity to handle uncertainty has been systematically removed. As ESPRIT 1217 has demonstrated, a human-centred system is less vulnerable to disturbance, displays less downtime and is significantly more productive[2].

These are, of course, the instrumental and economic considerations. Of more long term significance are the issues of human dignity, motivation and the form, nature and role of science and technology as the most extraordinary millennium in human history edges precariously to the dawn of the 21st century.

The present book is rich in the issues covered, embracing such topics as industrial culture, diversity, anthropocentric systems, political and economic interdependence and much, much more. Fortunately, these are not merely viewed from the usual Eurocentric standpoint. They may be said to represent a truly global view and reflect the insights of researchers, academics and policy makers from Japan, China, Soviet Union, Sri Lanka, Australia, Hungary, France, Belgium, Denmark, Italy and England.

Given Japan's pivotal role in the world economy, the growing interest in human-centred systems in that country is an important boost for ideas which started out so unpretentiously at Lucas Aerospace in 1973. Tokyo Keizai University is to be thanked for deciding to celebrate its 90th anniversary by hosting this event in Tokyo.

Notes

[1] Cooley, M (1987) *Architect or Bee?* Chatto & Windus, London. Japanese edition ISBN 4-275-01329-8 COO 36 P329E Tokyo 1989.

[2] Rosenbrock, H H (1990) *Machines with a Purpose*, Oxford University Press, London. Forthcoming Japanese edition Tokyo 1992.

Contents

Contributors

Achille Ardigò
Faculty of Political Sciences, and Director of CERDSI (Centre for Research and Documentation on Computer and Society), Department of Sociology, University of Bologna, Italy

Richard J. Badham
STS (Science and Technology Studies) Department, University of Wollongong, Australia

Mike Cooley
Independent Researcher and Centre Chairman of SEAKE Centre (Social and Educational Applications of Knowledge Engineering), Brighton Polytechnic, England

Alexander Dynkin
Institute of World Economy and International Relations, Academy of Sciences, (former) USSR

Marc Giget
Euroconsult, France

Karamjit S. Gill
SEAKE Centre (Social and Educational Applications of Knowledge Engineering), Brighton Polytechnic, England, and the editor of Springer-Verlag's Artificial Intelligence and Society book series

Satinder P. Gill
Department of Experimental Psychology, University of Cambridge, England

Susantha Goonatilake
Research Department, Peoples Bank of Sri Lanka

András Hernádi
Institute for World Economics, Hungarian Academy of Sciences

Peter Holden
Innovation and Technology Assessment Unit, School of Policy Studies, Cranfield Institute of Technology, England

Yukiko Inoue
Graduate School of Business Administration, Tokyo Keizai University, Japan

Yuji Masuda
Faculty of Business Administration, Tokyo Keizai University, Japan

Kuniko Mochimaru
Graduate School of Business Administration, Tokyo Keizai University, Japan

Massimo Negrotti
Methodology of Human Sciences IMES, University of Urbino, Italy

Takao Nuki
Musashi University, Japan

Hiroshi Okumura
Department of Economics at Ryukoku University, Japan

Lauge Baungaard Rasmussen
Institute for Social Sciences, Technical University of Denmark

Yoshihiro Sato
Research Institute of System Science, NTT Data Communications Systems Corporation, Tokyo, Japan

Fumihiko Satofuka
Sagami Women's University, Japan

David Smith
SEAKE Centre (Social and Educational Applications of Knowledge Engineering), Brighton Polytechnic, England

Kazuo Takeuchi
Tokyo Keizai University, Japan

Poul Tøttrup
Department of Production Technology, Danish Technological Institute

Werner Wobbe
Directorate-General for Science, Research and Development
Joint Research Centre, Programme FAST, EC Belgium

Valery K. Zaitsev
Japan Economic Section, Institute of World Economy and International Relations, Academy of Science, (former) USSR

Feng Zhao-Kui
Institute of Japanese Studies, Academy of Social Sciences, China

Section I:

Industrial Cultures and Technology Development

The Use of New Technologies in the Development of Social Citizenship and of the Welfare State

Achille Ardigò

The Welfare State has to cope with the consequences of new technologies[1]. The term Welfare State refers to the allocation by the State of public funds and services in order to assure to every citizen as a political right and not as charity, a minimum standard of income, nutrition, health, housing and education.

New technologies are driving forces of economic developments but their social effects are not pre-determined, or, at least, not uni-directional. In my opinion, it is rather a relation of double contingency - and double-dealing as well. This is the standpoint from which I would like to explore aspects of the *nexus between the Welfare State and new technologies*.[2]

Recent economic developments - driven and supported by an unprecedented flow of technological innovations - have undoubtedly contributed to an attack on a certain model of a centralised, corporatively or bureaucratically managed Welfare State.

New technologies, on the other hand, can facilitate the resolution of the crisis occurring in the centralistic Welfare State and at the same time give support for maintaining the rights of citizens, by enabling better provision of civic tele-information. Besides, new technologies can become precious instruments for the human promotion of disabled and aged people.

It is unquestionable that the social policies of States have been affected by economic and social development brought about through the influence exerted since the 1970s by computers and the telematic revolution.

A change in the composition of the labour forces has taken place, resulting in more professional and less manual workers, more self-employed and less employees.

The electronic multi-media revolution has turned the news about the most significant events in the life of mankind into mass information, conveyed in real time, although screened by economic, intellectual and political powers.

The social and cultural development of peoples, together with ageing of the population and growing access to world news, have intensified the demand for services and welfare provision from governments, as well as strengthening the wish for a better standard of living, including environmental safety. Increased pressure for public expenditure - mainly for pensions, health services and more recently for environmental policies - has critically burdened public budgets, even in countries with a strong private welfare market, like the USA. The electronic revolution has encouraged a better understanding of the modern states' social duties - especially towards the sick, the needy, and the disabled.

New technologies can, however, make an instrumental contribution in other directions, promoting divergent patterns in the structure of welfare states. An alternative to the perspective of increasing centralization and differentiation of welfare services along market lines is its diametrical opposite, a non-hierarchical, network-like welfare society.

Let's take health policies as an example. In this field, the model largely diffused in Western Europe involves constant expenditure to maintain bureaucratic systems with large - often iatrogenic - public residential institutions such as hospitals, large shelters and nursing homes.

A policy of reorganisation should instead be implemented through a detailed system of local day-care centres, supporting, as far as possible, home care.

The effectiveness of this strategy, surely, will depend mainly on the kind of practitioners and administrators involved in deciding upon and monitoring the new opportunities available in the restructured Welfare States. But it will also depend on the use-model incorporated in the new technologies as regards not only commercial, political and production-related targets, but also a use-model that takes in social and communicative ends.

In the model of the Welfare State in force up to the 1970s the leading actors were the policy-makers of the States and the leaders of trade unions. It can be said that a relevant portion of the expansive strategies of welfare states was then carried out either within the centralised industrial negotiations or within the system of political exchange.

But now, this model has to be reshaped.

A decentralised network of operators, both professional and voluntary, along with more active and widespread participation by citizens and users, will be the necessary key to a flexible and pluralistic - no longer centralistic - reorganization of the Welfare State, towards the *Welfare Society*.

In this context it was noted by Ralph Dahrendorf that there is no scarcity of authors, both from the Right and the Left, who claim the need for less state enforcement of civil rights, less state and more personal participation in the social policy-making process implemented through small and informal networks, self-help groups and voluntary associations etc.[3]

What is new here about this already widespread restructuring trend, is that active participation by individuals and by small groups in the management and control of social policies can benefit greatly by appropriate use of the new technologies.

This non-reductive reshufflement of Welfare States, supposes three reciprocally interrelated prerequisites:

a) *A decentralised participation of the private sector*, both profit and non-profit, in the management of welfare services, with special attention paid to home care for the sick and disabled and a rebirth of solidary communication at the neighbourhood level, beyond the nuclear family solidarity and as a substitute for so many elders living alone;

b) *The social use of new technologies to support civic participation and for an improved social control of welfare services*. All this should take place within a global network of territorial services, from hospital to general practitioners to an improved involvement of relatives, neighbours and local associations of health and social care;

c) *An increase in the number of opportunities for the citizens and the users to have a timely access to computer-provided information on Welfare States operations*. The actual right to civic information should be reshaped as a parallel result, so that a real right to information could be enshrined.

The above mentioned prerequisites contribute to the outline of *a new dimension of social citizenship*. The notion of social citizenship was introduced by the English sociologist T. H. Marshall. In many books published from 1950 Marshall[4] analysed the contents and forms of the exercise of the rights of citizenship as an evolutionary concept, wherein three parts or elements could be distinguished: the civil, the political and the social. (It shall be noted that very recently, many sociologists, even Marxists, in Anglo-saxon contests, have revisited the ideas of Marshall[5]).

Social citizenship, in Marshall's view, was to be considered distinct from, while being well related to, those rights practically necessary for the freedom of the individual, including the right to own private property and to take part in the exercise of political power. According to Marshall, social citizenship is the right/duty to the entitlements, historically specific, essential to living. Every citizen is entitled to these guarantees which are their due by the state. The entitlements cover 'the whole range going from a minimum of welfare and security to the opportunity to live the life of a civil person, according to the standards which obtain society'[6].

Social citizenship means a new way to recognize the right to public assistance for those who lose - temporarily or permanently - their self-sufficiency. For their part, they are expected to make efforts subsequently to regain their autonomy as active citizens and not only as workers or owners.

It must be said that these basic concepts of universalism within the public welfare procedures supposed a sufficiently solidary society, with a strong cultural integration inside and some civil economic and political values in common. After almost three decades, what comes to the sociologists' attention in the advanced societies is, rather, a decrease in solidarity along with *an increase in cultural heterogeneity*, oriented in cosmopolian directions.

There is a growing gap between generations. A gap even, in the same towns, between residential areas of younger couples and families and that of elders. Couples and families, still founding cores of any solidarity, show inconsistencies. The family shrinks, in the number of its components, and the proportion of singles grows. Social classes, centres of solidarity inside the social strata of workers, professionals and entrepreneurs, have broken up and weakened in their capacity to function as cultural and organisational references. The remarkable fall of the typical class conflict in industrial societies is counterbalanced by more and more visible signs of anomie and involuntary loneliness.

The advanced western societies become increasingly cosmopolitan, partly as a result of large-scale international migrations. Ethnic, cultural, linguistic and religious heterogeneity grows at an unprecedented rate. Serious difficulties arise in social communication - especially in the poorer areas of western cities - and xenophobia and intolerance phenomena are experienced.

Hence, given the difficulties and drawbacks to the social integration of people in the western world, *the need for more and newer communication opportunities* is strongly felt. The domain of social citizenship requires to be up-dated and extended, beyond though still in the wake, of Marshall's theory. Ralph Dahrendorf for instance in a brief reflection on

Marshall's work, refers to the 'gradual expansion of citizenship from legal to the political and social sphere' and adds that this process 'is still unfinished' as 'new dimensions of citizenship' may be discovered[7]. In my opinion, one of the new dimensions is exactly the social use of new technologies for welfare services and networks. More generally the reshuffle of the Welfare State would certainly benefit greatly from the socially enlightened use of the new telecommunication technologies.

After all, telecommunication support is already effective for the couple, family and kinship. Through the closely-woven telephone network, 'tele-family' can bridge geographical gaps. Where family and kinship are inadequate or absent, local telephone networks - equipped with specially conceived devices - are increasingly managed by private aid organisations, with professional and volunteer personnel. There are medical emergency networks, providing for medical aid and for the psychological relief of the old and disabled living alone, who may rely on emergency services, working day and night. There are *help line networks* for those who are threatened or suffer from physical violence at home or outside, and for other desperate people.

In spite of the commonplace difficulties of innovation, networks of telematic medicine are slowly being set up. They assure long-distance medical check ups for ill people living at home or in small villages, where specialist medical assistance is unavailable or not adequate. Computerized booking services for diagnostic tests and specialist examinations, provided by public health institutions, are being installed in many social centres in different cities. It is expected that these services will be extended also to GPs consulting room with tele-networks from GPs rooms to hospitals.

The new technologies are entering the different countries, in different ways and levels, in the restructuring of the Welfare State as a *mixed, three dimensional system* - public, private non-profit, and private commercial. A system is coming which will follow less and less, at least as far as the management is concerned, a hierarchical model. It is, rather, a pluralist model.

The challenge for these networks and their distinctiveness will be matching the systemic and formal communication coming from the institutional structures of welfare states, with the informal, talkative, spontaneous communications in natural language. In these networks, formal knowledge should be capable of communicating with 'tacit knowledge' (in Wittgenstein's view[8]), and vice versa. To achieve this goal, well-skilled cultural intermediaries are required, in order to deal effectively with the ethnic-cultural growing plurality of users of the Welfare State, be they actual or potential. This explains the need for a double-contingency telematic network pattern, removed where possible far from bureaucratic asymmetries.

At least six *network contexts* can be located and suggested as the recipients of welfare investment redistribution in the new technologies. These are meant to be a due social-targeted share of the economic benefits coming from publicly-financed R&D accomplishments of commercial and production activities.

a) *Socio-sanitary telephonic networks*, such as booking and information services for diagnostic tests, specialist's examinations and similar, which would reduce waiting times at the windows of the local health services.

b) *Medical telephonic networks (tele-networks)*, equipped with proper computer devices, mainly directed to linking the houses of people at risk (chronic invalids, convalescents, severely disabled, terminally sick people etc.), with hospitals and other specialised health institutions, especially for *medical tele-monitoring* [9];

c) *Tele-networks of civic information centres*, ensuring the citizen's right to information and also the access of immigrants to useful data and know-how procedures. These services could be devoted to social, participative and decentralized use for the purposes of communication and democratic social control[10].

d) *Social tele-networks*, public, private or mixed, for flows of formal and informal communication that follow up, control and reassure individuals, especially non self-sufficient people, living alone or within small households (of which members are often absent), and people who are afraid to be alone at home, especially at night. The same or other tele-networks would offering shopping and home banking opportunities and a better access to information sources on social allocations of goods and services.

e) *The use of information technology by social workers* in tele-networks, up to the use of AI utilities in form of expert support systems, 'for use by social workers to facilitate the gathering, dissemination and exchange of practice and development information'[11] and for professional training. In this field, the now prototypical development and use of artificial intelligence applied to welfare services would contribute not only to rationalizing social policy but also to developing techniques more flexible means of problem solving. The CERDSI centre of the Department of Sociology at Bologna University, will in the next few months carry out experiments in three local sanitary units in Northern Italy, with a computer based system incorporating elements of AI, to support decision making in public home care Assistance for the elderly. The ANCHISE project (that is the name of the AI programme) has as its purpose to support local decision makers (public social workers of the Italian NHS) in deciding between different types of assistance, following selective criteria and information on elderly people.

f) *Other non medical tele-networks to link outside centres with handicapped people*, who are confined to their homes have limited mobility and/or limited telecommunicating ability. Handicapped persons may be equipped, as a new Welfare State allocation, with interconnected displays, computers and special telephone sets - and other types of new informatics-based prostheses and devices - in order to overcome social marginalization and to promote employment of disabled people[12]. Reference is made here to new technologies/informatics supported prosthesis allowing the blind, the deaf-mute, the audio-injured, the spastic and others disabled in their speaking and motor capacities, to have access to computers and telematic systems for information, for social and interpersonal communications, for education and job purposes.

In this direction there is a growing commitment by national telephone companies in the development of special services for the disabled. During this year 1990, the SIP, the main Italian telephone company largely owned by the state, has undertaken an ambitious project called 'Insieme' ('together'). The experiment which was started in an average town in central Italy, Macerate, aims at testing a system providing many services for the deaf-mute, the disabled and the sick - either for home care or for hospitalized people. It is a sociological experiment that will be followed and studied by CERDSI (the Centre for sociological studies and researches on new technologies of Bologna University, Italy).

Also, there have to be mentioned new computational technologies, especially connected with musical, artistic and expressive purposes which may be used by experts, educators, and relatives in the interpersonal and interactive processes, for stimulating the curiosity, and the creativity of the mentally and physically disabled.

A number of crucial issues arise at the micro-social level as sociologists involve themselves in prototypical experiments with new technologies and applied devices for the handicapped.

Three of these issues are briefly outlined as follows :

a) There are specific aspects of the relation between the disabled and the 'intelligent' machines. New professions could be shaped out, as anticipated by the results of initial tests carried out on small groups of voice-disabled, equipped with technologically advanced tools such as graphic synthesizers of spoken words. These anticipations help to foresee the demand of the general market, also in other fields like telecommunications.

b) We should not forget however, that the same relation disabled/intelligent machine can bring about disfunctional effects. There is the risk that the computational medium ends up by reducing the communication - especially the expressive one - between 'normal' and disabled people. What should have been an authentic and intersubjective communication is then discharged among computer displays, in the classrooms, the home and the offices. How can this outcome be avoided?

c) Where is the right equilibrium point - especially during the formative phase of the disabled - between the intersubjective play, the affective life on one hand and on the other, the education into instrumental rationality heightened by computer programming?

Of course, the initial use of these networks is often prototypical, sectional and separated - even within the same administration unit in the same town. Those first stages must have to be superceded in spite of opposition and difficulties, towards an integrated, open and multipurpose system.

And now a concluding remark on the relational nexus between a new model of welfare states and the new technologies.

The development of a flexible decentralized and widespread management of opportunities and services within the Welfare State by new technologies, particularly for the handicapped, and other weak groups of citizens or poor immigrants, is a specific political and socio-cultural problem. It is a new chapter of the Welfare State that must be opened; from the *Welfare State* towards the *Welfare Society*.

Cultural and political steps need to be taken at the national and international level, in order to guarantee and control a balanced propagation of these opportunities, especially for certain vulnerable groups in the population.

Notes

1 With new technology, I mean especially telecommunications and computers and many appliances of micro-electronics, in a growing bundle of interconnecting tools and services.

2 See: Dery, D (1981) *Computers in Welfare: The mis-match* Sage p. Beverly Hills

3 Dahrendorf, R (1987) *Fragmente eines neuen Liberalismus* Deutsche Verlags-Anstalt GmbH, Stuttgart, see chapt. 15.

4 Of Marshall, T H, see: (1950) *Citizenship and Social Class and other essays* Cambridge University Press; (1963) *Sociology at the crossroads and Other Essays* Heineman, London; (1973) *Class, Citizenship and Social Development* Westport, Connecticut, Greenwood Press; (1975) *Social Policy in the Twentieth Century* London Hutchinson, 4th ed.; (1981) *The Right to Welfare and Other Essays* London Heineman.

5 For a recent leftist review of Marshall's theory, see: Giddens, A (1982)*Class division, class conflict and citizenship's rights*, in his *Profiles and Critiques and Social Theory* London Macmillan; Barbalet, J M (1988) *Citizenship: rights, struggle and class inequality. Concepts in social sciences* Open University Press, esp. pp. 16-33, 64-79, 81-96, 108-111.

6 Ibidem, see introduction.

7 Dahrendorf, R (1973) A personal vote of thanks. In *'British Journal of Sociology'*, 24, 4, see pp. 410-411.

8 Wittgenstein, L (1953) *Philosophical Investigations* Oxford, see Paragraph 202. On 'tacit knowledge' in the present context, see also: Kjell S. Johannesen, Rule Following and Tacit Knowledge, In *AI & Society*, oct.-dec. 1988. In the same issue, see Satinder P Gill *On two AI Traditions*.

9 See: Mandil, S H (1988) Innovations and the Applications of Informatics in the health sector. In *Contacts* March/June 1988; Dept. of Health and Social Security (GB) (1985) *Information technology and FPCs Family Practitioner and Community Health Services* DHSS, NHS Information Technology Branch, London; Andersen, A (1985) *Results of a study into the data required to support the exchange of information between family practitioner services and hospital and community health services* London, DHSS 2v.

10 On this issue, see : Bergqvist, S-R (1978) *Information and Communication about municipal affairs: the use of various means for informing citizens to facilitate their participation* Council of Europe; see also Filley, C, & Sutherland, M (1985) *Computers, Communication and the Community in Action* Scottish Community Education Council, Edinburgh, 50pp

11 See Forrest, J & Williams, S (1987) *New Technology and Information Exchange in Social Services* Policy Studies Institute, London

12 See Odonnel, S (1986) *Software for the handicapped: a community programme success* Programme in Practice Newsletter no. 4, Sheffield; Sandhu J (1987) *Information technology and the employment of disabled people* In *Employment Gazette* London 1987

Dec. Vol 95, n12; Gates. E *My Computer works by sucking and nodding* In The Independent 12 June 1989; Marsh, P (1981) Technology for the Disabled. In *New Scientist* 26Nov, 1981; Office of Technology Assessment report (1982) *Technology and Handicapped people* US Congress, Washington DC; A.S.P.H.I. (1987) *Lo sviluppo professionale degli handicappati nel campo dell'informatica* Bologna, Borghigiani.

Working Women in Japan - After Enforcement of the Equal Opportunity in Employment Law

Yukiko Inoue

Introduction

Here in Japan three years have passed since the Law of Equal Opportunity in Employment came into force on April 1st, 1986. This Law is formally termed as *the law respecting the improvement of the welfare of women workers, including the guarantee of equal opportunity and treatment between men and women in employment.*

During the ten years which includes *International Women's Year,* and the *United Nations Decade for Women,* both the government and private organisations in all the countries of the world grappled with women's problems and made efforts to improve the situation of women. At length, the Law of Equal Opportunity in Employment (hereinafter referred to as the 'EOE Law or 'the Law') came into existence on June 1st, 1985 and was enforced in the next year as stated above. Have women, by virtue of the EOE Law, become able to make use of their abilities in the place of work? By working as hard as men, could women get job satisfaction? The author would like to find out the answers to such questions by considering how working women's situations changed following the three years since the EOE Law came into force.

The first section will review the background to the EOE Law, and the second section will outline the Law. As regards the outline of it, the author will take up part of the Law: *Promotion of Equal Opportunity and Treatment between Men and Women in Employment.* The author will give a full account of this article in the third section of this paper. In the third section the author will give consideration to working women's situations with a survey report titled *Rengo-seisaku Shiryo* (Survey for Policy Planning). This survey was carried out by the National Federation of Private Sector's Workers' Unions of Japan (hereinafter referred to as RENGO) in 1988. Also the author will compare RENGO's survey report with two other ones entitled *Fujin Rodo Mondai Kenkyo* (Studies on Working Women's Problems) and *Tokyo No Fujin Rodo Jijyo* (Women Workers' Situations in Tokyo).

The former was carried out by the study Group on Working Women's Problems in 1988 and the latter was carried out by the Labour Economic Affairs Bureau of the Tokyo Metropolitan Government in 1988.

The Background to the Law of Equal Opportunity in Employment

The power of women's movements all over the world moved the United Nations at long last, and it developed *International Women's Year,* and tbe *United Nations Decade for Women.* In this section the author will discuss the United Nations' initiatives and the Women's movements in Japan briefly.

The United Nations and Women's Movements in Japan

Equality between men and women has been a basic idea in the activities of the United Nations such as *the Charter of the United Nations* and the *Declaration on the Illumination of Forms of Discrimination Against Women.* Also the idea effect that maintaining world peace can be accomplished by the spreading of a Democratic Egalitarian Society, in which there is no discrimination because of race, creed and sex, has been running throughout the history of the United Nations. As stated before, in the background of the EOE Law there was a series of movements - for instance, International Women's Year in 1975 and a lot of women's liberation movements stemming from *the Universal Declaration of Human Rights of the United Nations* which was adopted on December 10th, 1948.

In the case of Japan, Ayako Oba says, both newly-organised groups and the existing women's groups became active. Most of all, the core movement of such women's groups was the one to promote the ratification of the *Convention on the illumination of All Forms of Discrimination Against Women* (1979). Until the Japanese government decided to sign the above-mentioned convention in 1980, not only strong solidarity among all the women's organisations but also constant pressure group movements were needed. Even so it took another five years to achieve the ratification of the Convention. During this stage, enforcing the enactment of the law respecting equal employment was difficult. Women in labour unions banded together, going beyond the organisational bounds of the General Council of Trade Unions of Japan, the Japanese Confederation of Labour and the like. However, the rumour had spread that managements were going to make a statement to the effect that they were against the establishment of the law[1].

To make matters worse, labour unions run by male leaders were reluctant to grapple with this problem together in an organised way. Under these circumstances, civil women's groups united with all the women in labour unions in spite of their different creeds and policies, and carried on demanding that the law be upheld by the government, every political party and the management unions. The demand for the law respecting equal opportunity in employment was a major goal of the Women Workers' Movement, and at the same time it was a broad women's movement involving housewives and all kinds of women workers[2].

The author heard that the movement was not only a struggle between labour and management, and between men and women, but also a struggle between women and women; that is, unskilled women workers and professional women workers. In short, the movement started with the acquisition of women's suffrage and went on to the ratification of the *Convention on the Illumination of All Forms of Discrimination Against Women,* and could just manage to be embodied in *the Law of Equal Opportunity in Employment.* It is true that this law came into force with much difficulty as did the acquisition of women's suffrage before the Second World War. As it is prescribed clearly in the 14th article of the constitution of Japan[3], equal opportunity in employment is a basic human right. However, in practice we cannot see any big change concerning women's working situations. The EOE Law was eventually enforced though, and the author will discuss this in the third section of the paper.

Outline of the Law of Equal Opportunity in Employment

The EOE Law consists of four chapters and thirty-five sections in all. The first chapter is *General Provisions*, the second, *Promotion of Equal Opportunity and Treatment Between Men and Women in Employment*, the third, *Measures Assisting the Employment of Women Workers*, and the fourth, *Miscellaneous Provisions*. The following sections extracted from the *Legislative Series* of the International Labour Office are only part of this Law, but they are the most important ones for women workers. The author will begin with the *Purpose* of the law, and follow up with *Basic aspects and Measures to be taken by employers*.

Purpose

Section 1
The purpose of this law is to promote equal opportunity and treatment between men and women in employment in accordance with the principle contained in the Constitution of Japan which ensures equality under the law, fosters measures for women workers including the development and improvement of their vocational abilities, the provision of assistance for their re-employment, and attempts to harmonize their working life with family life; and thereby to improve the welfare and status of women workers.

Basic aspects

Section 2
In view of the fact that women workers contribute to the development of the economy and society and at the same time have a significant role to play as a member of the family in nursing children - who will be the mainstay of the future - the basic objectives of the improvement of the welfare of women workers as stated in this Law are to enable them to obtain job satisfaction by making effective use of their abilities, with due respect for their maternity but without discriminatory treatment on the basis of sex, and to achieve harmony between their working life and family life.

Section 3
Women workers shall, in awareness that they are members of the business community, endeavour, by their own initiative, to develop, improve and make good use of their abilities in their working life.

Measures to Be Taken by Employers

Section 7: Recruitment and hiring
With regard to the recruitment and hiring of workers, employers shall endeavour to give women as equal opportunity as men.

Section 8: Assignment and promotion
With regard to the assignment and promotion of workers, employers shall endeavour to treat women workers as equally as men workers.

Section 9: Vocational training
With regard to the vocational training for the acquisition of basic skills necessary for workers to perform their duties as provided by ordinance of the Ministry of Labour, employers shall not discriminate against women workers by reason of their being a woman.

Section 10: Fringe benefits
With regard to loans for housing and other similar fringe benefits as provided by ordinance of the Ministry of Labour, employers shall not discriminate against women workers by reason of their being a woman.

Section 11: Compulsory retirement age, resignation and dismissal
(1) With regard to the compulsory retirement age and dismissal of workers, employers shall not discriminate against women workers by reason of their being a woman. (2) Employers shall not stipulate marriage, pregnancy or childbirth as a reason for resignation of women workers. (3) Employers shall not dismiss women workers by reason of marriage, pregnancy or childbirth or for taking the leave stipulated in section 65, subsection (1) or (2), of the Labour Standards Law.

The above-mentioned sections are extremely important parts of the Law for women workers. It is a matter for regret, however, that this Law is so weak that employers will be able to escape punishment or get off comparatively lightly even if they violate the Law. It is a problem indeed and something must be done about this without delay.

The Situations of Working Women After the EOE Law

With the three survey reports of RENGO, the *Labour Economic Affairs Bureau of Tokyo* (LEABT) and the Study Group on Working Women's Problems (SGWWP), the author would like to give careful consideration to the situations of working women as well as comparing the three reports.

The RENGO Survey

This survey was carried out in the form of a questionnaire survey including open-ended questions in 1988. 1,778 questionnaires were distributed and the collection rate was 61.1 percent. The types of industry surveyed were: (1) Food, (2) Textile, (3) Paper and pulp, (4) Oil, (5) Chemistry, (6) Wholesale and retail, (7) Finance, insurance and real estate and (8) Information services and the like. In general, what areas have changed by virtue of the Law? So far as shown in Table 1, a remarkable change is found in *recruitment and hiring* because 33.7 percent of women workers surveyed answered that their companies have improved in the area of recruitment and hiring. And those of the *wage system* (21.7%) and *mandatory retirement age* (16.9%) follow it. The areas concerning *child care leave* (12.4%), *wage system* (12.0%) and *promotion* (9.6%) are being negotiated to make improvements between labour and management.

Table 1

Areas (A) improved, or (B) negotiating with the management by enforcement of the EOE Law (% shows the percentage of total companies)

	(A)	(B)
Recruitment and hiring	33.7%	3.1%
Wage system	21.7%	12.0%
Mandatory retirement age	16.9%	3.9%
Vocational training	14.0%	5.1%
Promotion	13.9%	9.6%
Child care leave	7.3%	12.4%
Re-employment system for women	5.4%	8.8%

Source: RENGO, Seisaku Shiryo (Survey for Policy Planning)
RENGO, 1989, p.30.

As regards *assignment and job-rotation*, it is clear that the range of women's work is expanding greatly and a large number of women are engaged in the same range of work as men. To take one example, a computer programmer or a system engineer, many women are working now in the information industries and the like. As for *promotion*, however, there is little chance for women workers to gain promotion in the same way as men. Especially in the manufacturing industries, women workers cannot be promoted with ease, although this remark does not exactly fit the service industries, and there are a large number of women who are playing an active part as managers in the field of retail, sales promotion and so forth. In the area of *vocational training*, many enterprises give the same training to both new men and women employees. However, they do not give any training to middle-aged women workers, or if they do, the training is different from that for middle-aged men workers.

Another thing we must pay attention to is that although discrepancy between retirement ages of men and women is prohibited by the EOE Law, and special measures for pregnancy or child care leave[4] are spreading little by little, there are still many women who resign by reason of marriage, pregnancy, or child-care. It is imagined that the two main causes for this are:

1) that nuclear families are increasing rapidly but there are few nurseries and nursing homes with reasonable fees, and
2) the lack of women's job consciousness, that is, low morale in the job.

Nearly eighty percent of all women workers are employed in medium and small sized enterprises, or small-scale enterprises where not even a union is organized. Generally child-care leave and in-house nurseries are available for workers in large enterprises. Therefore, pregnant women who work for small-sized enterprises can not but resign because, as stated above, they are not allowed to be absent from the company for a period.

Now let us look at the area of *re-employment for women.* The 25th section of the Law talks about propagation of special measures for re-employment as follows:

(1) In respect of women who have resigned by reason of pregnancy or childbirth or in order to take care of children, employers shall endeavour, where necessary, to give special consideration to the recruitment and hiring of such women who, at the time of resignation, expressed the wish to be re-employed by the previous employer,
(2) The state is to endeavour to give assistance, including necessary advice or guidance to employers, with a view to promoting the propagation of the special measures of re-employment stated in the preceding subsection.

Table 2

Re-employment system for women (The 25th section)

Has been introduced	13.1%
Under negotiation with company	9.0%
Under consideration in union	13.8%
No plan to introduce for the present	56.8%
Others and no answers	7.3%

Source: RENGO, ibid., p.58.

According to Table 2, only 13.1 percent of the enterprises have introduced *the re-employment system for women* so far, and 9.0 percent of women workers surveyed found that it is under negotiation between union and management. Comparing each industry, "the retail, finance and service industries" are introducing this system the most. However, it is most important for women to keep on working however hard the work is. This survey concludes that bringing equal employment between men and women into place is a unions' responsibility, too. Therefore labour unions must make every effort to realize that employers should be punished if they break the EOE Law.

The survey of the Labour Economic Affairs Bureau

This survey was carried out in 1988 and 2,500 women workers responded to the questionnaires. The type of industries surveyed are almost the same as those of RENGO. As shown in Table 3, we can understand that sex discrimination in employment was improved to a certain extent by virtue of the EOE Law. The area of *improvement of work rules* was the one improved most (34.8% of total companies) and *improvement of wage system for women* (26.5%), *improvement of promotion for women* (21.5%) and *expansion of placement for women* (21.0%) follow it in that order.

In this survey too, the business which underwent remarkable improvements after the enforcement of the EOE Law are service industries such as finance, insurance and retail and the like as was found in the RENGO survey.

As a whole in this survey, 74.3 percent of women workers surveyed found that there exists discrimination against women in their workplace, especially discrimination against women in *wages* (72.5%), *promotion* (69.3%) and *placement* (41.2%).

Table 3

Improvement regarding sex discrimination after the enforcement of the EOE Law (%)

Improvement of work rules	34.8%
Expansion of hiring for women	18.2%
Expansion of placement for women	21.0%
Improvement of wage system for women	26.5%
Improvement of promotion for women	21.5%
Increase of women managers	9.9%
Improvement of discriminatory retirement age	12.0%

(Answers are multiple).

Source: TOKYO NO FUJIN RODO JIJYO (Women workers Situations in Tokyo), Labour Economic Affairs Bureau, 1989, p.42.

The survey of the Study Group on Working Women's Problems

This survey was carried out in 1988. It distributed 585 questionnaires and the collection rate was 54.0 percent. The types of industries surveyed are manufacturing, service industries, finance and insurance industries, wholesale and retail and the like. Table 4 depicts that 49.7 percent of the workers surveyed found discrimination against women in recruitment and 56.6 percent of them found there is discrimination against women in placement which has not changed so far. Regarding promotion, only 3.1 percent of them found that equality of promotion prospects improved, and 10.6 percent of them found vocational training between men and women is the same. To the author's surprise, however, with regard to retirement age 76.4 percent of the women workers surveyed answered mandatory retirement age has been the same between men and women since before the EOE LAW. A further surprise is that 31.1 percent of them answered that a re-employment system is available. As far as the author can see, regarding this percentage, it seems that future prospects of a re-employment system is not so bleak.

Table 4

How the situations of working women have changed after the enforcement of the EOE Law (%)

Discrimination against women in recruitment and hiring exists as ever	49.7%
Discrimination against women in placement has not changed at all	56.6%
Vocational training between men and women is the same	10.6%
Retirement age of men and women is the same since before the EOE Law	74.4%
Promotion for women has improved equally with men	3.1%

Source: Extracted from FUJIN RODO MONDAI KENKYU<15> (Studies on working women's problems), the study Group on Working Women's Problems, 1989, p.45. (Answers are multiple).

Comparison of Three Survey Reports

As far as we know, as stated above, there are three survey reports on the Law of Equal Opportunity in Employment for Working Women in the past three years. Though the author mainly used RENGO's report, the others are also interesting and it is meaningful to compare the three surveys.

The author would like to state that the average age of the women workers surveyed by SGWWP is 38.4 years old. According to this survey,as a whole, the situations of women workers did not change so much by virtue of the EOE Law. Most of all, the wage system is the worst and 70.2 percent of women workers answered that their wages have not changed at all. It is said generally that the wages of women workers in Japan are only half as much as those of men. In the case of RENGO, we do not know the average age of women surveyed, but 72.5 percent of them pointed out the discrepancy in wages between men and women, and 36.7 percent of them said that their abilities are not brought into full play in their working places.

In the case of LEABT (average age of women surveyed is 29.8 years old), only 21.7 percent of them found the wage system was improved after the Law as shown in table 1. Needless to say, the length of women workers' service and promotion speed are different from those of men workers. We can say, however, that discrimination against women in wages is one of the biggest discrepancies between men and women and it is as high as ever. Although this does not apply to specialisms, engineers and managers, most of the women are working as clerks and the like. So the discrepancy between men and women in wages has not improved on one the hand and the discrepancy between women and women in wages shows a tendency to increase on the other hand. This is the real situation of women workers' wages today in Japan.

Now the author would like to look at the area of placement in the three surveys. Only 7.2 percent of women workers surveyed by RENGO found that placement and job-rotation improved after the enforcement of the Law. In the case of LEABT, 21.0 percent of them found that placement for women expanded after the EOE Law (Table 3), and 56.6 percent of them in the survey of SGWWP found that discrimination against women in placement exists (Table 4). Judging from these three reports together, the Law has not had an effect on the placement of women workers, and lots of women are still assisting men's work. The RENGO report says, however, in the retail business, and financial and service industries women workers are well represented in almost all workplaces, and that they are playing an active part in them as well. Even in the other businesses, it may be true that not only will Equal Employment advance more, and women will work throughout their life like men do, but also women will make full use of their abilities in every job. In addition to this, professional women, or women starting their own businesses, will increase in the near future. However the author also anticipate that temporary women workers will increase in number because it is convenient for both companies and women themselves. The fact is that the number of working women has been increasing rapidly in recent years[5], and the number of women who continue to work will increase on the one hand and women who work temporarily will decrease on the other. Anyway, in this ageing society, the author is sure that women will become a more and more important workforce for companies.

Finally, it is said that the advancement of child-care leave and a re-employment system are two of the most important areas to enable women to keep on working. But as far as this goes, there is no sensible difference between the surveys of RENGO and that of SGWWP. Though the form of questionnaires differ from each other, the former says that the rate of improvement in child-care is 7.3 percent and that of a re-employment system for women is 5.4 percent respectively (Table 1). In the latter, however, about one-third of women workers surveyed answered that they can already use such systems. And as to the retail business, financial and service industries, the rate of companies with a re-employment system is 37.0 percent. In the case of LEABT, about half of the women workers say they want the advancement of child-care and a re-employment system the most, but it is not clear whether such systems themselves exist in the companies surveyed at present. So it is risky for us to come to conclusions from the three reports. We can say, however, that not only the advancement of employment but also the expansion of social facilities such as nursery and nursing homes must be improved as soon as possible. Also, every woman working for any business or industry has the right to receive such benefits.

Concluding Remarks

The Law has not had a great effect on working women's situations so far. The effect of this Law is still weak in the matter of improvement of the discrepancy between men and women in employment, and as RENGO's report says, labour unions should make greater efforts to change this. However, only three years have passed since the EOE Law was brought into force, so we must take a long-term view of it, too.

We may conclude that one of the most important things is women's attitude towards work, that is, women's morale. The author would like to say once again that women have had to make an effort to continue to work whatever difficulties they have come across. In addition, in order to achieve promotion as fast, and increase wages at the same rate as men, they should improve their skills and develop their abilities in every possible way. Finally, most of all, women continuing within employment is indispensable.

References

Akamatsu, Ryoko (1985) *The Law of Equal Opportunity in Employment and Revised Labour Standard Law*, Japan Institute of Labour.

Ministry of Labour (1989) *Handbook of the Law of Equal Opportunity in Employment and Labour Standard Law* (Revised Edition), Japan Institute for Women's Employment

Oba, Ayako (1988) *Prehistory of the law of Equal Opportunity in Employment: Notes of women's labour history after the war*, Mirai-sha

International Labour Office (1986) *Legislative Series*: 1985-Jap.1, Int. Labour organisation

RENGO (1989) *RENGO-Seisaku shiryo* No.49 (Survey for Policy Planning) the survey report concerning women workers situations after the enforcement of the Law of Equal Opportunity in Employment, RENGO (the National Federation of Private Sector's Workers' Unions of Japan)

The Study Group on Working Women's problems (1989) *FUJIN RODO MONDAI KENKYU* (No.15), the Study Group on Working Women's Problems

Labour Economic Affairs bureau of Tokyo (1989) *TOKYO NO FUJIN RODO JIJYO* (Women workers' situations in Tokyo), Labour Economic Affairs Bureau of Tokyo Metropolitan Government)

Ministry of Labour (1989) *FUJIN RODO NO JITSUJYO* (the situations of women workers), Printing Bureau of Ministry of Finance

Notes

[1] Ayako Oba, Prehistory of the Law of Equal Opportunity in Employment, 1988, p.19.

[2] Ayako Oba, ibid., pp.19-20

[3] The 14th article of the constitution of Japan: All of the people are equal under the law and there shall be no discrimination in political, economic or social relations because of race, creed, sex, social status or family.

[4] The 28th section (propagation of child-care leave): Employers shall endeavour to take measures for child care for women workers. That is, women workers with children should be allowed, upon request, to absent themselves for a certain period to look after their children.

[5] According to FUJIN RODO NO JITSUJYO (The Situations of Women Workers, 1989, Women's Bureau of Ministry of Labour), the population of women workers in 1988 was 24.73 millions and both an increase in number as well as an increase in percentage was higher than those of men workers.

Adjustment in Foreign Assignment: A Comparative Study of Japanese and American Expatriates in Australia

Kuniko Mochimaru

Introduction

This paper compares the adjustment problems in foreign assignment between American and Japanese expatriates in Australia. The analysis will use a method with strong relations to Robinsons' classification into six roles of the expatriates. Hereby, the analysis will discuss which of the two countries' expatriates is the most successful regarding the following issues: members of the parent company; residents of the host country; citizens of the home country; members of the subsidiary; members of a specialists' group; and members of families or equivalents.

Part I: Literature Review

This part consists of four sections. The first section will define the adjustment in relation to the foreign assignment. The remaining three sections will present issues appearing in each adjustment aspect as defined in the first section.

Definition of Central Notions

Some cultural shock at the beginning of foreign assignment and during the following adjustment is unavoidable for the expatriates (Harris & Moran 1983). Desatnick and Bennett (1977) discuss adjustment as follows:

> *To assume a high-level managerial position in a foreign environment requires physical, mental and psychological adaptability. It may be necessary for the expatriate to change eating habits, social protocol, possibly even life style. He cannot remain aloof from his new environment and the local people: he must assimilate their culture, traditions, habits and customs, and not spend all his spare time in a ghetto of his own countrymen.*

This attitude and behaviour, however, will solve only part of the adjustment problems of the expatriates because adjustment has many aspects. There are several roles an expatriate can play. Robinson (1984) categorized six:-

(1) representative of the parent company
(2) administrator of the subsidiary company
(3) resident of the host country
(4) citizen of the home country of the parent company or of an other country
(5) member of a specialist group
(6) member of a family or an equivalent

Here, these roles are re-defined, in accordance with the current personnel situation of the Japanese companies, that is, sending more non-executive expatriates with Japanese nationality:

(1) employee from the parent company
(2) member of the subsidiary company
(3) resident of the host country
(4) citizen of the home country
(5) member of a specialist group
(6) member of a family or an equivalent (Mochimaru 1989).

Adjustment should be considered from the four aspects in which the problems occur; *internal* and *external* features both *abroad* and *after repatriation*. Adjustment occurs both inside and outside of the organisation. The important fact is that adjustment is not unidirectional. It also occurs when the expatriates go back to their home country. Consequently assimilation is not the ultimate answer to the adjustment problem. An expatriate with an attitude and behaviour as described by Desatnick and Bennett, therefore, is accepted as a member of the subsidiary and as a resident of the host country, but not as an employee from the parent company nor as a citizen of the home country. The latter roles require the expatriate to maintain his culture, traditions, habits, and customs. The dilemma with the former roles will be reconciled by the expatriate's efforts to understand the local language, culture, customs, etc.

The following three sections will discuss the issues of adjustment which appear in each one of the four aspects named here; first, *internal adjustment abroad*, second, *external adjustment abroad*, and third, both *internal and external readjustment*. As a result, the discussion will clarify to which role(s) of the expatriate each aspect is/are related.

Internal Adjustment Abroad

Both Japanese and American literature claim the importance of people's adjustment to working in overseas assignment (Fields & Shaw 1985, Iwanai 1987). The job situation, the working conditions, and human relations in general are the main factors in internal adjustment abroad.

Job Situation

The job responsibility and the relation with the parent company will be discussed.

Job responsibility

Job responsibility relates to three of the six roles; *employee from the parent company* (1), *member of the subsidiary company* (2), and *member of a specialists' group* (5).

The job responsibility is higher abroad than at home as the number of management staff is fewer in foreign subsidiaries and also as the middle manager at home has to operate also as general manager abroad. Some expatriates cannot adjust to such expanding responsibility. It is also pointed out that the responsibility is higher in American companies than in Japanese companies, which give American expatriates more power of operation in the subsidiary company (Ishida 1985). The high responsibility abroad results in more problems on their return to their home country.

Relation with the parent company

The relation with the parent company is not only related to the role of *employee from the parent company* (1) but also as *member of the subsidiary company* (2).

The expatriates often claim that parent companies do not understand the situation of the foreign subsidiaries. By providing information on domestic affairs through newsletters, company publications, etc., the parent companies can, however, ease the anxiety of the expatriates about estrangement from the parent company. As this problem is more closely related to the repatriation, it will be further discussed later, namely as a readjustment problem along with the job responsibility problem.

Working Condition

The working condition includes compensation and amount of work.

Compensation

The following discussion indicates that the compensation is related to all six roles of the expatriate.

The discussion by American scholars is based on the fact that the remuneration of expatriates is high. Robinson (1984) defends the necessity of high salary for expatriates as follows:

a) persuading them to leave their home country
b) keeping the same level of living standard as in their home country, when they are assigned to countries where the living standard is lower than that in their home country
c) expenses for home-leave
d) for educating their children

e) discharging their social duty to their family, to their friends, etc., at home.

Edstrom & Galbraith (1977) limit the application of a high salary policy only toward the transfers for filling a position. Transfers for management development should be motivated by promotion, not by compensation. Japanese companies are, on the other hand, advised to pay more to their expatriates for a higher living standard, which corresponds to the recent discussion on compensation for domestic corporate employees (The Nihon Keizai Shimbun Sept.13, 1987).

Working hours and benefits

The working hours, and benefits such as paid holidays, and the home-leave system, all relate to four of the six roles; *employee from the parent company* (1), *member of the subsidiary company* (2), *citizen of the home country* (4), and *member of a family or an equivalent* (6).

Ishida (1985) pointed out that the Japanese expatriates work very long hours and take few paid holidays. The home-leave system of the Japanese companies is also insufficient. Although more than 90% of Japanese companies offer a home-leave system to their expatriates, the standard ratio is only one home-leave in three years. The duration is ten days to one month, and the expenses for trip are paid for a specified destination, Japan (Kinoshita 1983).

American companies, on the other hand, offer home-leave to their expatriates every or every other year. The companies pay expenses for the trip without a specified destination or give flight tickets (Stone 1986). As the purpose of the home-leave system is to prevent the expatriates from losing the living pattern of the home country, payment for the home leave in cases where it is not taken or used for a trip to places other than the home country will cause a complaint from the host national employees (Noer 1979).

Human relations with the host country nationals (HCNs)

Human relations are related to three roles: *employee from the parent company* (1), *member of the subsidiary company* (2), and *resident of the host country* (4).

Factors which influence the relationship between expatriates and HCNs are the difference in perception of working (Inamura 1980), expatriates' high standard of living (Zeira 1975), languages (Almaney 1974), etc.

Japanese expatriates are criticized for:-

a) working too long after regular working hours
b) entertaining executives visiting from the parent company
c) excluding the HCNs (Inamura 1980).

Also, the HCNs suspect that the expatriates make very important decisions without them.

Both the expatriates and the parent company need to improve their efforts to get better relations with the HCNs. In order to understand the local culture, the expatriates should accept HCNs and themselves as they are, with patient and sensible mind and without unrealistic expectations, and should accept the foreign assignment as the challenge of participating in a cross-cultural experience (Harris & Moran 1983). Sakamoto (1975) recommends expatriates to foster personal friendship with the HCNs.

The expatriate executive's willingness and ability to learn a native language or appreciate the subtleties of a foreign non-verbal system will give a better impression to the HCNs (Almaney 1974).

The parent company can help the expatriates to have good relations with HCNs by explaining to the HCNs the personnel policies on expatriates in order to avoid their misunderstanding the expatriates gorgeous life (Zeira 1975).

External Adjustment Abroad

The potential failure of foreign assignment is indeed family-related; factors such as a wife's adjustment difficulties or children's educational problems are often, rather than expatriates' job factors, potential pitfalls. This is, of course, unless family life is directly influenced by the work, such as long or frequent business trips (Labovits 1977). The adjustment of wives and of children, and the expatriates and their families relations with the local community and with their home country will be discussed in this section.

All these problems except the relations with the home country, are related to three roles; resident of the host country (3), citizen of the home country (4), and member of a family or equivalent (6). The relation with the home country is related to two of six roles; citizen of the home country (4) and member of a family or equivalent (6).

Adjustment by wives

The adjustment by wives depends on the relationship between husband and wife commonly seen in the home country of the expatriate. Some of the oriental practices such as excluding wives from the social affairs cause some complaints by American wives (Baker 1976), who are used to be with husbands at parties. On the other hand, Japanese wives try to prevent isolation by keeping contact with other wives (The Asahi Shimbun May 13.1987). They are accustomed to stay home when their husbands have business dinners after work. It should be noted, though, the friction among the Japanese wives is sometimes greater than that between the Japanese and the HCNs (Ichimura 1980, Inamura 1980, Oosawa 1987).

Adjustment by children

School-age children might be exposed to various problems such as limited freedom of mobility in the area with bad security, fewer friends, fewer places to play, less amusement than at home, and relations with servants or maids (Inamura 1980). This is especially true in developing countries where the expatriates and their families live in the areas only for foreigners and sometimes also for upper class HCNs segregated from the ordinary HCNs.

Japanese schools seem to have become a 'must' in overseas Japanese communities, which should be well accepted in the sense that the Japanese children abroad are given the same educational opportunity as those in Japan. However it will be a problem if it means the children are enclosed in competitive-only circumstances as in Japan (The Asahi Shimbun Jan.18.1988). The Japanese school in relation to the local community will be discussed in the following section.

A rather new factor is the pre-school age children. The influence of the overseas experience at pre-school age in the process of development should not be ignored (Morimune 1987). This is becoming a more important problem as it is the recent trend that the younger employees are sent abroad.

American schools have not been discussed in the literature on expatriates. The Americans including the scholars of business never seem to have doubted the existence of American schools in foreign soil. This is a problem to be examined further.

Relations with the local community

Japanese groupism is criticized when discussing the relation to the local community. A Japanese scholar advised not to make a 'Japanese Village' (Ueno 1987).

In addition to the Japanese school, the *Juka* or the cram school which started business abroad made the Japanese children further isolate from the local community (The Nihon Keizai Shimbun Feb.13.1988), as they go to such schools after they come home from the local school, while the local children usually are playing. The Japanese children miss the chances of making personal contact with the local children. The Nihonjin Kai or the Japanese Society which has been believed to be a bad example of Japanese groupism, on the other hand, is expected not only to provide the Japanese win the opportunities to make friends among other Japanese for psychological relief but also to make better relationships with the local community (Hamada 1977).

Relations with the home country

The expatriates and their families wish to keep contact with their geographically separated kin and friends (Hamada 1977). Such feeling is strongest in cases with aged parents in the home country (The Asahi Shimbun Aug.21.1987). From this point, too, the home-leave system is inevitable and should be evaluated and reformed.

Readjustment

There are both internal and external adjustment problems back in the home country as well as in foreign countries. Many scholars have discussed readjustment problems which are more difficult than the those concerning adjustment to foreign environments (Sieveking, Anchor, & Marson 1981).

Internal Readjustment

Ignoring the repatriates' adjustment problems might mean their resignation, which will be a loss of time and money for the companies (Management Review, August 1982, pp53-54), and which might produce the trend of rejection among the employees toward the foreign assignment (Harvey 1982).

The job, compensation, and human relations will be discussed in this section.

Job

Changes in organisation and in state of authority, under-evaluated overseas' experience, the different value system, and the technology lag cause the readjustment problems. They are all related to all roles of expatriates except member of a family or an equivalent.

1. Changes in organisation and in state of authority

The organisational change at home (Freemantle 1978) such as the change of the executives, the transfer of sponsors, the promotion of friends or of co-workers will embarrass the repatriates as they do not know how to deal with such a new situation (Smith 1975).

The state of authority is a big problem for expatriate managers, especially top managers. The expatriates who played a big roll in a small stage abroad will return as a small fish in a big pond at home (Noer 1987). They feel reverse culture shock because of the loss of social and professional status (Desatnick & Bennett 1977).

2. Evaluation of overseas' experience

No or insufficient use of the overseas experience of repatriates have been repeatedly criticized (Noer 1978, Labovitz 1977).

3. Other problems

Adjustment problems will also occur from the difference of working tempo or of the efficiency, or from the alienation at the business negotiation (Inamura 1980). The lag in technology is serious for engineers, typically repatriates from the developing countries (Edstrom & Galbraith 1977).

Some countermeasures should be taken to facilitate their readjustment. Information on domestic projects and on the home country will keep the expatriates identifying themselves with the parent company and with the home country. The participation in the domestic related projects is one of the useful means to link the expatriates and the parent companies (Desatnick & Bennett 1977).

Compensation

The problem exists in promotion and in remuneration. They are related to all roles except *member of a family or an equivalent*.

Delay of promotion is the problem and is one of the demerits of foreign assignment (Labovits 1977.)

The remuneration problem is also the status problem. The top managers especially feel the social decline on the fall of total income (Desatnick & Bennett 1977), and by the loss of the status and privilege as executives (Yasumuro 1983). It is important to give them incentives by paying a good salary (Desatnick & Bennet 1977). The problem of remuneration from the point of individual asset will be discussed as the external readjustment problem.

Human Relations

Human relations are related to two roles; employee from the parent company (1) and member of the subsidiary (2).

The lack of adjustment in human relations at the parent company is caused by the pressure from the upper and lower colleagues or by friction with them (Desatnick & Bennett 1977). A comparative study was done on this problem among Japanese repatriates in various industries. Those in trading companies, banks, security companies, where more Japanese are working at the overseas offices, found the readjustment easier than those in manufacturing industries with fewer expatriates abroad (Sussman 1985).

External Readjustment

The compensation, the readjustment of wives and children, and repatriation plan will be discussed.

Compensation

Compensation is related to four roles; *employee from the parent company* (1), *member of the subsidiary company* (2), *citizen of the home country* (4), and *member of a family or an equivalent* (6).

Some of the demerits of foreign assignment are found in the unstable life and economic loss (Toyo Keizai Tokei Shimpo, January 1981). The expatriate may miss the chance of owning their house. Therefore the asset planning should accompany the relocation counselling (Management Review August 1982, pp.53-54).

Readjustment of wives

Wives' readjustment is related to three roles; *resident of the host country* (3), *citizen of the home country* (4), and *member of a family or an equivalent* (6).

Various external problems will come to the wives upon repatriation. The wife has to listen to her husband's complaints about his new job as well as to cope with her new life (Labovitz 1977) and help their children's readjustment (Inamura 1980).

Japanese wives have to mingle with the daily problems such as human relations with neighbours, relatives, friends, etc.; small houses; expensive goods; small town; and narrow streets (Inamura 1980); and to adjust to the traditional role as housewife from the couple-oriented social life abroad (Kumar, Steimann, & Nagamura 1984). The American wife also hates returning to a suburban housewife again (Smith 1975). The wives' experience should also be referred by the future expatriates (Baker 1976).

Readjustment of children

Childrens' readjustment relates to the roles of *resident of the host country* (3), *citizen of the home country* (4), and the *member of a family or an equivalent* (6).

Teenage children may show stronger attachment to their friends made during their overseas life (Smith 1975). Uprooting may become an obstacle for smooth repatriation (Desatnick & Bennett 1977).

The readjustment problem in Japan most discussed is of children's education. There are two other reasons for difficulty of readjustment in addition to the severe entrance examination to universities and colleges: the lack of tolerance in Japanese human relations (Inamura 1980) and the less considered personal benefit in the Japanese society (Kumar, Steimann, & Nagamura 1984).

Repatriation plan

Several countermeasures to deal with the whole readjustment problem are suggested. The fundamentals for solving the readjustment problem is that the companies recognize the existence of the problem (Sussman 1985). Then the planned repatriation and the care for the expatriates follows. The most advanced counteract is to make a repatriation plan at the selection (Sieveking, Anchor, & Marson 1981) or before departure to overseas (Harvey 1982).

The God Father System (Noer 1978) is suggested; and the Reentry Boss system of NEC (Romu Jijo, November 1987, pp.55-59) is the practice of its theory. The system includes the detailed relocation and regularly revised repatriation plan to make long-range counteracts toward readjustment. If such a long-range plan is impossible to make, earlier (three months to one year prior) announcement of repatriation will reduce the difficulty of readjustment (Labovitz 1977). The expatriates will then have more time to prepare for going back both physically and mentally.

The duration (average: three years) should be prolonged as this is in the midst of adjustment period from the points of work efficiency, productivity and mutual understanding with the host country's community (Inamura 1980). The timing of returning should also be influenced by the children's education, as much as possible (Labovits 1977).

This chapter revealed the relationship between issues of adjustment and the roles of expatriates as shown in figure 1. However, the literature examined here do not compare the American and Japanese expatriates on the issues of adjustment in the same environment for accurate comparison. This is why the empirical study was planned.

The following figure illustrates the above discussed issues of (re-)adjustment of expatriates.

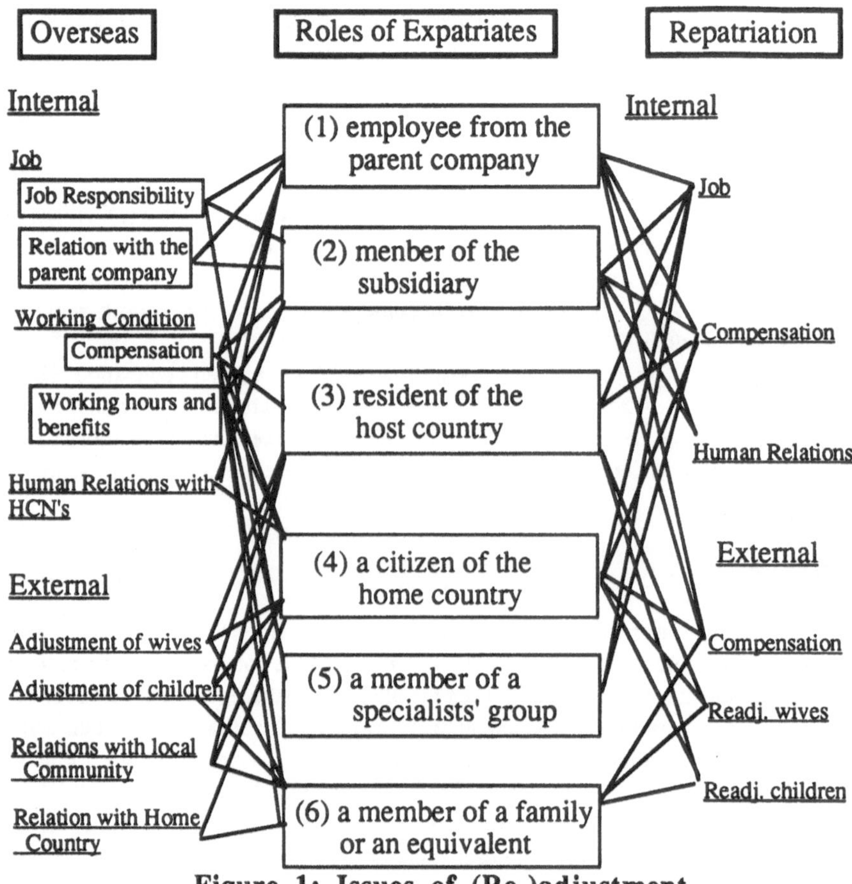

Figure 1: Issues of (Re-)adjustment

Part II: Methodology

Two Japanese (A, B) and two American (C, D) companies were selected for this study. All four companies have subsidiaries in Australia and operate production lines in auto industry. Companies A and C are located in the Melbourne area, companies B and D, in the Sydney area.

The Japanese auto industry recently started production in advanced industrial nations. The American auto industry, on the other hand, has a long history of overseas operation since before the World War II. The comparison of the behaviours of the both national expatriates in Australia will be a hint for the Japanese companies' personnel management of the expatriates assigned to other advanced nations such as the United States.

Three types of questionnaires were prepared. One was for the personnel department of each parent company (Type 1). The second was for the expatriates and their families in Australia (Type 2). The third was for the Australian employees (Type 3).

The interviews followed the Type 1 questionnaires for the Japanese companies. As for company D an expatriate executive in the Tokyo office (regional centre) answered the Type 1 questionnaire. The type 2 questionnaires were answered by nine out of fifteen expatriates of company A, two out of three of company B, three out of sixteen of company C, and one out of one of Company D. Three Japanese expatriates including the two heads, one American expatriate of Company D, were interviewed. The Type 3 questionnaires were answered by eight out of 2,450 (150 managers) of Company A, two out of five department managers of Company B, three out of 11,700 of company C, and by one out of thirteen department managers of Company D. Most of the respondents are managers. Five Australian employees including the Executive Vice-President and the personnel manager of Company B were interviewed. The profile of the four companies are shown in Table 1.

Table 1: Profiles of Companies A, B, C and D (August 1988)

Companies	A	B	C	D
First Overseas Production	1957	1974	1904	1930's
Domestic Employees	55,000*	6,000*	180,000	140,000
Overseas Employees	20,000*	570*	170,000	20,000*
Overseas Sales				
(vs Whole sales)(%)	33	<10	20*	30*
Employees in Australia	2,450	140	11,700	366
Expatriates in Australia	15	4	16	1

* Approximate figure.

Part III: Results and Discussion

The adjustment abroad and readjustment both internally and externally will be analysed on Japanese and American expatriates. The analysis will determine in which roles Japanese and American expatriates are successful respectively.

Internal adjustment Abroad

The job, the working condition, and the human relations will be analysed.

Job

Though both the Japanese and the Americans found their job enjoyable, they have some problems. The Japanese problems are caused by either social (difference of culture or of working pace, structure of society) or individual (language, way of supporting the Australian managers) factors. However, they are doing well, even with some problems, as an expatriate of Company A answers that job has always some problems and no one answered 'not enjoying work with problems.' One of the American expatriates has got the problem that the parent company did not show the clear goal. The rest three Americans found their job very enjoyable. This difference may have been caused by the length of stay in Australia; the former, two years, the latter, less than one year.

Job responsibility

The job responsibility was not compared between Japanese and Americans as the positions of those expatriates were different. A Japanese respondent appreciated the broadened managerial span due to small size of the subsidiary which gives him better chances to participate in the company (subsidiary)'s managerial strategy.

Relation with the parent company

Relation with the parent company is compared by three factors: communication with the parent company, participation to the decision making in the parent company, and the job performance evaluation. The latter two factors were not included in the literature review.

The Japanese were found to be more successful from this aspect.

1. Communication with the parent company

All four companies use similar means for communicating with the parent companies. Main means are in-house newsletter, daily communication, and business trip to the parent company. At company A, the information on the parent company is also brought by the people of the parent company on their business trip to Australia.

2. Participation to the decision making in the parent company

Three out of nine expatriates of Company A answered 75-100% participation rate. 50-75% is the highest of other companies. As the participation rate depends on the expatriates' position or how much of the domestic decision is related to Australian operation. It can be said that the expatriates and the parent company tie stronger in Company A than the ones in the other companies or that the original positions of Company A expatriates at home were higher than the other expatriates.

3. Job performance evaluation

The differences in control of the expatriates show how the performance of the expatriates is evaluated.

There is a decisive difference between the Japanese and American companies evaluation system. The American expatriates are evaluated by their Australian supervisors. The Japanese evaluation system is varied. As the expatriates of Company A are all advisers to the

Australian managers, the Japanese are evaluated by those managers and the supervisors in the parent company. The executive manager of Company B is evaluated by his supervisor in Japan, and the non-manager, by his Australian supervisor. Such evaluation system relying much on the parent company might produce the trend of looking to the parent company. On the other hand, the American system or the system of Company B as for the non-managerial expatriate might create problems at repatriation as they may tend to look to the subsidiary to get better evaluation and hence will not be ready for their repatriation.

Working condition

Four companies are compared on compensation and on amount of work.

Compensation

It could not be determined who is more successful on this issue. Most of the Japanese and American expatriates are satisfied with their present salaries. Although seven out of nine Japanese expatriates did not expect high salaries from the beginning, one of seven found his salary insufficient. Two expatriates of Company B complained. One of them had to pay more than expected for the furniture, electric appliances, and so on for settling into a new life in Australia. This situation happened because of the parent company's less experience of expatriates management and will not happen to more recent expatriates as the parent company, after a little trouble with this expatriate on this problem, established the rule of compensating more at the beginning of the transfer for smooth settling. His age (early 30's) might be another factor for the complaint as his generation was brought up in richer circumstances than the previous generation (40's) where other Japanese expatriates belong to. This case reminds the author of the claim for higher salary discussed in Chapter II. The reasons of the other expatriates' complaint are not known. The remuneration of Company B might be lower because of its smaller size and of its short history of overseas operation.

An expatriate of Company C complained as one of his motives for accepting overseas assignment was high salary and also probably as he worked longer than the other two. Though there is no data to prove that his salary is insufficient, he is privileged to be paid much according to Edstrom & Galbraith (1977), as he was probably transferred to fill a position as designer.

Working hours and benefits

Working hours and benefits are compared by the working hours per day, paid holidays, and the home-leave system. On these issues the Americans were found to be much more successful than the Japanese.

The medium of the working hour of the Japanese (ten hours per day) is the same as the longest of the Americans. Though this result does not immediately mean that the Japanese work longer than the Americans as the sample is small, the references by the Australian employees on excessive working of Japanese expatriates support this fact.

The decisive difference is seen in the use rate of paid holidays. Most of the Japanese take 0 to 50% (almost half of them: 0 to 25%) paid holidays while Americans take 75 to 100% of them (100% by three out of four).

The negative view of the Australian employees toward such too much commitment to the companies will be discussed as a human relation problem.

The home-leave systems of companies A and D bear some problems, though they are completely opposite. Company A provides the home-leave once in three years and claims it enough as the people of the home country including the expatriates' parent(s) visit them. However it is doubtful if the expatriates would feel the whole environment of their home country so as to avoid the reverse cultural shock when they come back home from abroad, by less frequent return trips. Company D does not designate how to use the expenses for home-leave paid to the expatriates. This might have caused the complaint of an Australian employee of Company D on the compensation of the expatriates. Neither of the companies understand the real meaning of the home-leave system.

Human relations with Australian Employees

Human relations with Australian employees are discussed from the points of the expatriates and the Australians, and of the personal friendship among employees. The relations between Australians and Japanese are mainly analysed. Fewer comment on the relation between Australians and Americans can mean less problem between them.

The expatriates' point of view

The relation with Australians is commented only by the Japanese. The Japanese expatriates appreciate the hard-working Vietnamese. The Japanese working ethics is still alive in Australia. It is also pointed out that Australians do not like to be trained by the Japanese while the Thai people, from an expatriate's experience, find it natural to be trained by the Japanese.

The Australian's point of view

The complaints of Australian employees toward the Japanese are on the excessive working of Japanese expatriates, and the complaints toward the Americans, their high salary and benefits.

On the contrary to the Japanese comment, it was found that Australians wanted the Japanese to teach them by taking them to Japan or by sending more Japanese to Australia in Company B. This is a particular case of lack of communication. Apart from this, there is a problem of value difference. The Australian employees are found to feel some difference of value in the Japanese excessive commitment to work or company as discussed previously, status of women, and the detailed planning. Most of them do not show their displeasure toward the Japanese expatriates because of these differences. However, one Australian feel uncomfortable about it and wrote that the Japanese did not like the fact that the Australians did not work such long hours. This story shows that the expatriates bear unrealistic expectations and that do not accept the HCNs as they are.

The complaint toward the American expatriates are on their high salary and housing. Though the parent company explains the personnel policy to the HCNs, the reasons of high salaries does not seem to be explained.

One courageous finding for the Japanese, who tend to weigh the ability of English too much, is that the Australians who have lived in the immigrants' society do not tend to judge the job ability of a person by the ability of English.

Personal friendship among employees

The question is whether it is accepted in Australia to make personal friendship among employees apart from the office. Australian employees answered in three ways: Yes, No, and Yes but with some condition. Most of the Australians in Japanese companies answer Yes or Yes with some condition while those in American companies, half, Yes and other half, No. In practice, almost all have experienced some kinds of personal friendship such as attending the wedding party, BBQ party, etc. This fact supports the Sakamoto's advice (1975) to promote personal friendship with the HCNs. The Japanese have long believed that the Western people do not have personal contact among the employees after work. This may affect the behaviour of the Japanese expatriates to try to limit their contact with HCNs only during the working hours. One condition should be added though, namely that such friendship does not affect the job in any way.

External Adjustment Abroad

Four companies are compared on the adjustment of wives and of children, and the relation with the community.

Adjustment of wives

Although most of the Japanese expatriates and their families answered that they were enjoying their life in Australia, the longer they live in Australia, the more they tend to complain. Unreliability toward Australians and the worry on repatriation are the main reasons of their complaint. Language is also a part of their problems. To keep their own language or to learn a new language is their concern. The friction among the Japanese wives was not their problem as there are fewer Japanese expatriates in one firm.

The Americans seem to be more comfortable. They express no complaint. This is probably because Australian language, culture, and customs are very similar to their own.

Adjustment of children

Although American children feel a little better than the Japanese children, the problems exist in both sides. As Australia belongs to the advanced countries, no such problems discussed on developing countries in Chapter II do not appear in the answer sheets.

The problem which annoys the Japanese children most is the severe entrance examination for colleges and Universities in Japan. Most of the Japanese children go to the Japanese school. Company A recognises the existence of the adjustment problem of pre-school children and presents the results of questionnaire on pre-schoolers' adjustment in one of their newsletters on education for their expatriates and the families.

As there is no American school in Australia, the American children go to a school with Australian children. Although English is the common language for the Americans and Australians, and the influence of American culture is much seen in Australian TV programs, the American children face some problems. The Australian school system is rather British where the children have to choose their future earlier, which does not identify with the American system where the children can choose it later.

Relation with the local community

Areas to live and the structure of friends will be discussed The Americans are more successful than the Japanese on the relation with the local community.

Areas to live

The Japanese tend to gather in the areas convenient for their children to go to the Japanese school. It should also be noted that some of the Japanese chose the areas with fewer Japanese avoiding the complicated human relations among Japanese.

The location of the subsidiary affects the adjustment of the expatriates. Melbourne has fewer Japanese (2,859) than Sydney (4,116). The Japanese society in Sydney does not seem to be active either in a way of attracting more Japanese or making a role as a bridge between Japanese expatriates and Australians. There exists no such society in Melbourne. The Japanese school is rather new in Melbourne (established in 1986.) These facts may help the Japanese in Melbourne more localised than those in Sydney if other conditions are the same. Americans scatter since their children go to Australian schools.

Structure of Friends

The big difference appears in the structure of friends. While 64.4% of Japanese friends are Japanese, 75.0% of American fiends are Australians. As there are much more Americans (67,600) (1987) than Japanese (9,063) (1987) in Australia, it is no wonder if the Americans had more American friends. Therefore, this figure shows that Americans mentally adjust to Australian life more successfully than the Japanese. It should be noted that one Japanese had no Australian friend and the other has no Japanese friends. These might be the causes of maladjustment. Some of them have third-country national friends through the hobby groups, which shows their global minds and the strong ties with the local community. Many of the friends were made through their children, which means the expatriates should expose their children to the host country community to make HCN friends. In this sense, the fact that the Japanese School in Sydney is open to Australians should be appreciated.

Relation with the home country

Relation with the home country was already discussed as internal problem relating to the home-leave system.

Readjustment

Japanese and Americans are compared on both internal and external readjustments. It should be noted that the discussion here had to be limited to the problems the respondents can predict before they go back to their home country.

Internal Readjustment

Readjustment will be analysed from the aspects of job, compensation, and some other problems.

Job

Changes in organisations and in state of authority, evaluation of overseas' experience, and technology lag were analysed. It was found that both Japanese and Americans worry toward the changes in organisations and in the state change of authority, and the evaluation of the expatriates' overseas experience.
1. Changes in organisations and in the state of authority

While the personal department had no worry on readjustment, the expatriates, especially younger ones do worry about repatriation. This is probably due to their insufficient human network in the parent company as they were transferred abroad before they have established it. The parent companies offer the expatriates information on the parent companies by several means such as the daily communication, the in-house newsletter, and the business trip to the parent company, which does not seem to be enough for assuring the smoothly readjustment for these young expatriates.

The expatriates worry that their responsibility or their power will diminish back at the parent company. These facts have been pointed out by the literature.

2. Evaluation of overseas experience

While the parent companies provide either domestic or overseas job for the repatriates, many expatriates wish to work abroad again. An expatriate of Company A pointed out, though, the expatriates' return to his previous job can activate the personnel along with the mid-year employment, domestic transfer, and the transfer between personnel and engineering departments.

3. Other problems

The technology lag also pointed out in literature are being overcomed by the advancement of telecommunication or of transportation.

Compensation

Americans are apparently more successful on this matter than the Japanese. Difference was found between Japanese and Americans concerning promotion as part of compensation.

The Americans expect promotion upon return while the Japanese have no such expectation, which seems to verify the discussion that the Japanese companies no longer send only elites abroad (The Nihon Keizai Shimbun Sept.13.1987). In Company C, the promotion at repatriation is determined before departure. If the expatriate is promoted during the overseas assignment, the position at repatriation will be re-examined at the parent company.

Unpredictable problems

The different value system, status change following the reduced remuneration, and human relations discussed in the literature are the problems that the respondents can not predict before repatriation.

External readjustment

Childrens' readjustment and repatriation program are the main problems. The remuneration and the wives readjustment cannot be predicted.

Childrens' readjustment

American children have smaller obstacle for repatriation. Although their common and most important problem is education, it appeared differently. While the severe entrance examinations to the university are in the background of the Japanese, the American problems occur in the lack of grades and of the academic year. These problems are caused by the difference of school system of respective countries. The Australian academic year starts in February. The American's starts in September.

The Americans might solve a part of the problem by adjusting the duration of assignment. On the other hand, in order to solve the Japanese problems, not only the educational system but also the mental structure of the Japanese parents and other adults who place too much stress on educational ability to judge the children's ability should be changed. It should be appreciated that Company A sends the leaflets on the current educational situations in Japan to their expatriates regularly to lesson their worry toward the children's education.

Repatriation plan

A Japanese expatriate worries about the housing back in the home country. It is understandable as the land price in Tokyo area in Japan is skyrocketing.

None of four companies answered that there is readjustment problem, nor they have any repatriation plan, which means the fundamentals for solving the readjustment problem lack in all four companies.

Part IV: Conclusion

Table 2 shows that which roles are related to each adjustment problem, who is more successful on each problem, and finally concludes who, Japanese or Americans, are more successful in each role. The Japanese expatriates are more successful as members of the parent company. The Japanese expatriates work longer with little expectations of high salary. They are more controlled by the parent company, though this may be due to the difference of the expatriates' state (Japanese; advisers in Company A only, Americans; managers). While the Americans expect promotion at repatriation, the Japanese do not.

The American expatriates are more successful as residents of the host country, citizens of the home country, and members of families or equivalents. The Americans whose parent companies understand the importance of the home-leave system go back to their home country more often than the Japanese. The main reason of success of the Americans as residents of the host country is probably that the Americans and the Australians share the same language, English. The Japanese has handicap in this sense.

Neither the Japanese nor the Americans are successful as member of the subsidiary.

Table 2. Comparison of succes of Japanese and American Expatriates in Relation to their Six Roles

ROLES / ITEMS	(1)	(2)	(3)	(4)	(5)	(6)	Who is more succesful
INTERNAL ADJUSTMENT							
Job Responsibility			Not compared				
Relation with parent company	A<J	A<J					A<<J
Compensation	A=J	A=J	A<J	A>J	A<J	A>J	A=J
Working hours and benefits	A>J	A>J		A>J		A>J	A>>J
Human relations with HCNs	A=J	A=J	A>J				A≥J
EXTERNAL ADJUSTMENT							
Wives			A>J	A>J		A>J	A>>J
Children			A>J	A>J		?	A≥J
Relation with Local Comm.				A>J		A>J	A>J
Relation with Home Country			A>J	A>J		A>J	A>J
INTERNAL READJUSTMENT							
Job	A=J	A=J	A=J	A=J	A=J		A=J
Compensation	A>J	A=J	A>J	A>J	?		A>J
Human Relations			unpredictable				
EXTERNAL READJUSTMENT							
Renumeration			Unpredictable				
Wives			Unpredictable				
Children			A>J	A>J		A>J	A>J
Repatriation Plan	A=J	A=J	A=J	A>J	A=J	A=J	A≥J

REFERENCES

Almaney, A (1974) Intercultural communication and the MNC executive. In *Columbia journal of world business*, winter 9:23-28.

Baker, J.C.(1976) Company policies and executives' wives abroad. In *Industrial relations*, October 15 (3): 343-348.

Desatnick, Robert L & Bennett, Margo L (1977) *Human resource Management in the Multinational Company*, Westmead, England, Gower press, Teakfield Limited.

Edstrom, A & Galbraith, J (1977) Alternative policies for international transfers of managers. In *Management International Review*, 17 (2): 11-23.

Fields, Mitchell W & Shaw, James B (1985) Transfers without trauma. *Personnel Journal*, May 64:58-63.

Freemantle, David (1978) Foreign assignment: a recruiters nightmare. *Personnel Management*, October 10:33-37.

Hamada, Reishi.(1977) Singapore, Malaysia no nipponjin shakai ni okeru nippon fujin no seikatsu to ishiki (A survey on the life consciousness of japanese women in Singapore and Malaysia - on its characters of japanese businessmen's wife). In *Rodo Mondai Keikyu* (Journal of Labor problems), November (6):41-51.

Harris, P R & Moran, R T (1983), Translated by International Relation Research Institute, Kokusai Shoka University. Ibunka Keieigaku (Managing Cultural Differences.) Tokyo: Perikan Sha.

Harvey, M C (1982) The other side of foreign assignments: dealing with the repatriation dilemma. In *Columbia journal of World Business*, Spring 17:53-59.

Ichimura, Shinichi (1980) *Nippon Kigyo in Asia* (Japanese Companies in Asia). Tokyo: Toyo Keizai Shinpo sha.

Inamura, Hiroshi (1980) *Nipponjin no Kaigai Futekio* (Japanese maladjustment abroad). Tokyo: NHK Books.

Ishida, Hideo (1985) *Nippon Kigyo no kokusai Jinji Kanri* (International Personnel Management of Japanese companies). Tokyo: Nippon Rodo Kyokai.

Iwanai, Ryoichi (1987) *Keizai Kyoshitsu* (Class of Economics). Nippon Keizai Shimbun.

Kinoshita, Akira (1983) *Jinteki Shigen no Kaigai Iten* (Overseas Transfer of Human Resources). Kyoto Keibun Sha.

Kumar, B, Steimann, H, & Nagamura, Y (1984) Doitsu no Nipponjin chuuzaiin. In *Rodo Mondai Kenkyu*, July 9:45-62.

Labovitz, G H (1977) Managing the personal side of the personnel move abroad, *SAM Advanced Management Journal*, Summer 42: 26-39.

Mochimaru, Kuniko (1989) Kaigai haken yoin no ikusei ni kansuru bunken kosatsu (Literature review on development, selection and training of expatriates). In *Tokeidai Ronso*, March 10:19-43.

Morimune, Takero (1987) Sanyo denki-yojiki kara chugakusei (Sanyo electric co-from toddlers to junior high students). In *Keieisha*, June:61-63.

Noer, D M Translated by Kinoshita, Akira (1978) *Kaigai Chuzaiin no Jinji Kanri* (Mutinational People Management.) Tokyo: Tokyo Nunoi Shuppan.

Osawa, Shuko (1987) Kaigai shinshutsu no uragawa de-ko no ibunka taio no byori gaku (Dark side of overseas operation-pathology on individual response toward the different culture). In *Honyaku No Sekai*, March:43-49.

Robinson, R D,Translated by Research Group of MNEs, chief: Irie, Itaro (1984) *Kihon Kokusai keiei Senryaku Ron* (Internationalisation of business.) Tokyo:Bunshindo.

Sakamoto, Yasumi (1975) Zaigai Kigyo no jinji romu kanri (Personnel administration in Japanese enterprises abroad). In *Nippon Rodo Kyokai Zasshi*, February: 2-15.

Sieveking, N, Anchor, K & Marson, R (1981) Selecting and preparing expatriate employees. In *Personnel Journal*, March 60:197-202.

Smith, L (1975) The hazards of coming home. In *Dun's Review*, October 106:71-73.

Stone, Raymond J (1986) Compensation pay and perks for overseas executives. *Personnel journal*, January 65:66-69.

Sussman, Nan M (1985) Problems after overseas assignment. In *The Japan Times*, September 25.

Ueno, Akira (1987) Kaigai shinshutsu ni seiko suru tameno 10 no joken (Ten conditions for successful overseas operation). In *Sekai Keizai Hyoron*, May: 55-60.

Yasumuro, Kenichi (1982) *Kokusai Keiei Kodo Ron* (International Management Behaviour). Rev.ED. Tokyo:Moriyama shoten.

Zeira, Y (1975) Overlooked Personnel problems: multinational corporation. In *Columbia Journal of World Business*, summer 10: 96-103.

Kaigai chuzaiin no shushi kessan (Income expense balance sheet of expatriates). Tokyo Keizai Tokei Geppo, January 1981:42.

Koyo shin jijo NO.25 Nippon Denki (New trend of personnel No.25 NEC). Romu Jijo, November 1987:55-59.

When the overseas executive comes home. In *Management review*, August 1982 712:53-54.

The Nihon Keizai Shimbun, Sept. 13, 1987 & Febr. 13, 1988.

The Asahi Shimbun May 13, 1987, August 21,1987, & January 18, 1988.

Koseisho Jinko Mondai Kenkyush. 1989. Jinko no Dodo Nippon to Sekai Jinko Tokei Shiryoshu 1988 (Trend of Population Japan and the World Dat book of Population Statistics 1988, Tokyo. Kosei Tokei Kyokai.

Cameron, Angus & Henwood , Belinda (1988) *The Australian Almanac & Book of Facts 1988*, Angus & Robertson publishers.

Intercorporate Relations in Japan

Hiroshi Okumura

Trade Relations between Corporations

We can identify four agents of trade: (1) individual consumers of familles, (2) corporations, (3) non-profit institutions (government, state, etc.), (4) foreign countries. When a corporation is paired as a common agent with all the other types of trade agent, there are four types of trading relationships formed: (1) corporation/individual consumer, (2) corporation/corporation (inter-corporate), (3) corporation/nonprofit institutions, and (4) corporation/foreign countries. Among the four types of trade relations, the second type, intercorporate, is conspicuously present in Japan. No accurate statistics exist to support this assertion, but many examples point to this conclusion. When compared to other nations, the volume of wholesale trade to retail trade is far larger in Japan. In the USA the difference between wholesale and retail trade is 1.6., 1.9 in the UK, 1.2 in France and 1.7 in West Germany, but 4.0 in Japan. Most wholesale trade is between corporations, whereas retail trade occurs mainly between corporations and individual consumers. This statistic indicates the large volume of intercorporate trading in Japan.

A good example of intercorporate trading can be found in the automobile industry. Japanese auto makers buy about 60 to 70% of their parts from other corporations. In contrast the big three American auto makers buy only 30 to 40% of their parts. Trade relations between Japanese steel makers and auto makers is another example of how intercorporate trading works in the automobile industry. Nippon Steel, for example, sells its steel only to the trading companies which in turn sell that steel to auto makers. In the USA, U.S.X. sells its steel directly to auto makers. Sogo Shosha, a general trading company, is unique, because it acts as an intermediary, it accelerates intercorporate trading.

There are several reasons why intercorporate trading is so extensive in Japan. Firstly, Japanese corporations are not as diversified as US corporations. They tend to specialize in certain narrow areas of production within the same trade industry. Sales may be undertaken by affiliate sales companies while the production process is subdivided and some areas of production are even sent to subcontractors. Second, heavy industry increases the amount of intercorporate trading because heavy industry has inclination of detour production in its character. In Japan the proportion of heavy industry to all other industries is very high. However the main reason for widespread intercorporate trading is that the capital relation between the corporations is very extensive and strong.

Certain characteristics distinguish intercorporate trading from trade relations between corporations and other trade agents. Firstly, Japanese corporations generally trade with only a small number of trading agents, so that he trading relationship between corporations and individual consumers differs from the intercorporate relationship because individual consumers are many because they are inclined to select and restrict the numbers of their trading partners. For example, the largest steel producer in the world, Nippon Steel, sells its products to only a dozen trading companies. Also Toyota Motors buys its automobile parts from only two hundred parts makers. Unlike Toyota, however, Daimler Benz, a German auto maker, buys its parts from over one thousand parts makers. So in these trading relations the corporations do not trade with many, unspecified partners, but only with a few, specified partners. This has very important implications because economic theory usually presupposes that the market mechanism works only between many, unspecified trading partners.

Another characteristic distinguishing intercorporate trading relations is the difference in principle from the so-called market mechanism. In classical economic theory the trading relations in the market mechanism is free and not specified. Corporations sell to or buy from who ever can offer the best price, quality or service. The 'invisible hand' works to determine the price and trading partner. Trade between corporations and individual consumers involves few corporations and a larger number of individual consumers. When large corporations dominate the market, it is called oligopoly. In the case of intercorporate relations, however, corporations trade face-to-face. Trading partners are specified and fixed, therefore their relationships became long term. This tendency in trade policy is especially seen in Japan.

In the market mechanism the price of goods is determined by the competition between unspecified sellers and buyers. In contrast, the exchange of goods between corporations is not. Intercorporate trade competition is different from market mechanism competition. Intercorporate trading corporations prefer, first, to select their trading partners and, second, to determine price. What works here is not the 'invisible hand', but the 'visible hand'. Suppose company A buys from company B. If company C offers a lower price than B, does A change from B to C? No: What happens is that company A will tell to company B that company C has offered a lower price, and B will offer the same lower price. Their trading relationship does not change; but the price will. In the case of intercorporate trading the first concern is the trading partner, after that the price.

In intercorporate trading competition between corporations is not excluded. It is, however, different from market mechanism competition. In the market there are many, unspecified trading partners that compete with each other. The price is determined by competition which, in turn determines the trading partners. Therefore there is always a competition for new partners. The hypothetical situation mentioned above may be re-examined, applying market mechanism principles of competition. Suppose there is a long term relation between company A and company B but company C is trying to enter the trading relationship. If company C offers a very low price and company B cannot offer the same price, or if C introduces a new product and B cannot offer the same product, company A will change its partner from company B to company C. Often these changes will occur when company A is trying to diversify its products or to apply a new method.

This type of intercorporate competition is a competition for trading partners. When a company tries to get new partners it will offer extraordinary price cuts for certain partners. The company does not consider the market. Some Japanese economists say this phenomenon is 'excessive competition', but this competition for new Partners is partial and short term, but heated.

In intercorporate trading we always find reciprocity: company A buys from company B if B buys from A. This reciprocal dealing is thought to be unfair trading because it restricts trade. Such reciprocal dealing is prohibited by law in the USA especially after the conglomerate merger movement in the 1960s. However, it is not very common in the USA. If company A and company B produce a single kind of product, A only sells to B, and cannot buy from B. Only when companies diversify their products is reciprocal dealing possible.

But the situation is different in Japan. The general trading companies (Sogo Shosha) serve as intermediaries for intercorporate trading. If company A sells to company B there is only a one way trade relation; there is no reciprocity. But if the general trading company is the intermediary then reciprocity can occur. Company A sells a product to company T (general trading company) and company T sells that product to company B. At the same time company A buys materials or parts from company T and company B buys from T. In this example there are two reciprocal deals, one between companies A and T and the other between companies B and T. Almost all big Japanese corporations trade with general trading companies, so there is in actuality a very large web of reciprocal dealings in Japan. Reciprocal deals are also very common within the enterprise groups. For example, in the Mitsubishi group, Mitsubishi Corporation (general trading company) serves as the intermediary for trade among all Mitsubishi group companies. This creates a dense reciprocal business. With reciprocal dealing the trading partners are fixed and a new entry is very difficult. Therefore if foreign companies try to enter into the Japanese market, they usually experience a lot of difficulty establishing reciprocal relations with Japanese companies.

Up to this point trade relations have been discussed only in terms of goods, but there are many trade relationships involving money, ie financial transactions. In this area the main bank system is peculiar to Japan. All Japanese corporations have a main bank, which is usually the largest lender to the member corporations. This is the so-called 'keiretsu yushi'. The main bank has the responsibility of caring for its member corporations. If a corporation goes bankrupt the main bank buys back bonds issued by the corporation. How do companies select their bank? Is it a result of competition? No. It was determined historically. The main bank for all Mitsubishi group companies, for example, is Mitsubishi Bank. The main bank comes first, afterwards the price (interest rate) is determined. The relationship between the main bank and the corporations is face-to-face. It is not a market relation. In this relationship there is also reciprocity. The main bank lends money to the corporation. The corporation in turn deposits the money in the main bank. It is an obliged account. This differs somewhat from compensating accounts in the USA.

There is competition among the banks to get the corporation to be its main bank. This competition is not centred on interest rates. Other factors are more important. It is not determined by the market mechanism either. The same situation is found between securities companies and corporations. All listed corporations have a lead manager securities company which underwrites the new issue of stock and bond. These relations are fixed and long-term.

Intercorporate relationships among manufactures, trading corporations, banks and securities companies are based on face-to-face relations not only in Japan, but in the USA and European countries too. In these countries, however, this relationship bears the character of a short or one time only relationship, not long term or fixed. Only in Japan can a long term, fixed relationship be found between corporations.

Capital Relations between the Corporations

All large Japanese corporations have many 'keiretsu' companies. These include affiliate, subordinate, and subcontracting companies. According to the Japanese Fair Trade Commission in 1986 the largest 100 non-financial corporations had 3,899 subsidiary companies (companies with over 50% of their issued stocks owned by the parent corporation) and 9,519 affiliate companies (companies with 10 to 50% of the issued stock owned by the parent corporation). This is an example of a vertical relationship in which the parent corporation control their 'keiretsu' companies unilaterally. The parent corporation owns their 'keiretsu' company stocks and sends directors unilaterally.

Large parent corporations also combine horizontally into 'enterprise groups'. Before the Second World War the 'zaibatsu' dominated the Japanese economy. Mitsui, Mitsubishi, Sumitomo and Yasuda were famous 'zaibatsu'. After the war, General MacArthur (Supreme Commander of the Allied Forces) dissolved the 'zaibatsu', but after the Korean War (1950-52) these 'zaibatsu' groups were reborn as the enterprise groups. Today we can find six major enterprise groups in Japan: Mitsubishi, Mitsui, Sumitomo, Fuyo, Daiichi-Kangyo and Sanwa groups. In these groups each member corporation owns other members stock and holds a position in the presidential club, where important information involving member corporations is discussed. These horizontal combinations are peculiar to the post war Japan. In fact the combined economic power of these six major enterprise groups is stronger than that of the pre-war 'zaibatsu'.

These vertical (keiretsu) and horizontal (enterprise group) relationships are based on capital relations, ie stock ownership. Many corporations own much of the stock in other corporations in Japan. Over 60% of all stocks listed are owned by corporations (including manufacturing, trading corporations, banks, etc.). This phenomenon is a conspicuous characteristic of the Japanese stock ownership. This phenomenon may be called the 'corporatisation' of stock ownership to distinguish it from the 'institutionalisation' of the stock ownership in the USA or UK. In the US and UK financial institutions (eg. pension funds, investment trusts, life insurance companies and trust departments of banks) own stocks as agents of others. But in Japan many corporations own stocks at their own risk in order to gain and maintain control of the companies. Institutional investors, however, own stock in order to earn profit for the principal. In Japan many corporations own a large amount of other corporation's stock to control them. This type of ownership makes the 'corporatisation' of stock ownership a phenomenon peculiar to Japan.

In the USA many big corporations own other corporation's stocks, but the amount of the stock is much less than that owned by the Japanese corporations. According to the Securities and Exchange Commission only about 15% of all stocks in USA are owned by corporations. Commercial banks do not own stocks because it is prohibited by the Bank Act. The Clayton Act also restricts the corporate ownership of stock. Therefore US corporations own much less stock than do Japanese corporations. The Japanese Anti-Monopoly Act restricts corporate stock ownership, but in reality the restrictio clause consists of dead words and has no effect. For example, over 50% of Japan Victor Company's stock is owned by Matsushita Electric. And Japan Victor Company competes in the same market with Matsushita Electric. Can you imagine a sinilar case in the USA?

According to the Japanese Fair Trade Commission the largest 100 non-financial corporations own the stocks valued at 7,9 trillion yen. In contrast their net capital value is 8.7 trillion yen. This value of stock, however is book value. If we calculate the market value, it amounts to 3 or 4 times the book value.

Why do Japanese corporations own so much stock? They own the stock to control other corporations. This point seems self-evident in the case of vertical intercorporate relations. But many scholars do not think control is an issue in the case of horizontal ownership. For example, in Mitsubishi group each member corporation owns only a small percentage of the other corporation's stock. They do not own enough stock to gain control. But if we look at the Mitsubishi group as a whole they own 20-30% of each corporation's stocks; so as a group they can control each other. This situation can be looked at as mutual control based on mutual ownership.

In the 1960s many Japanese corporations were afraid of being taken over by foreign corporations and tried to stabilize their own stock. Toyota motors, the pioneer, endeavoured to stabilize its stock by asking related banks, trading companies, steel companies and sub contractor companies to buy Toyota Motor's stock to defend it from a possible takeover. Toyota Motors stabilized over 50% of its stock in a short time. Many corporations followed Toyota's example. As a result of this practice, there are no take-overs in Japan; in contrast take-overs are an epidemic in the USA and UK. With the stabilization of stock, the corporation asks related corporations to own its stock. They buy stock to control each other, and once they buy stock they usually do not sell the stock they own.

Mutual stock holdings among corporations also leads to personal relationships, particularly to interlocking directorates. There are many cases of interlocking directors in Japan. But this practice is not nearly as common in Japan as it is in the USA. Besides interlocking directorships, relationships are formed among members of the president's clubs. Here the personal relationships are particularly evident.

Capital exchanges and the personal relationships based on them are long term and fixed. Some businessmen say this indicates a 'blood relationship' between the corporations, but we must not confuse the interpersonal with the intercorporate relation. It is important to distinguish between the corporation and the individual person. In interpersonal relations there is emotion; people love and hate, therefore the relationship is capable of change. But in intercorporate relations there is no emotion; the relationship is long term and fixed. In the enterprise group, many corporations have been linked with each other for over 100 years. It is important to notice that these long term, fixed relationships are always based on capital relations, ie stock ownership.

Principle of Intercorporate Relation

Generally speaking, economic theory presupposes the existence of market mechanisms that coordinate actions among individual people where there are many, unspecified producers and consumers. Economists think the price is determined by supply and demand in the market place. In this presupposition the subjects of economic conduct are many unspecified individuals. This leaves the market free and the price mechanism is the principle of all transactions.

At the end of the 19th century, big corporations emerged as the main producers. Accordingly economists advanced a theory of monopoly and oliogopoly. These theories, however, presuppose that producers are monopolists or oliogopolistic and that the consumers are a large number of unspecified individuals. They discuss only the number of producers, and do not take seriously the idea that corporations are different from individuals. They think that corporate conduct is identical to individual conduct. Hence they concluded that corporations and individuals would act according to the same principles in the same market. Their main concern is only whether the number of producers are few or many.

Criticizing these theories, Oliver Williamson advocated the transaction cost theory in his book *Market and Hierarchies*. Following this theory Japanese professor A. Goto wrote that the principle of intercorporate relation is neither market nor hierarchy, but the intermediate area between the two[1]. He thinks that if transaction costs increase, the corporation (=hierarchy) will internalize the costs, but if administrative costs are too high, corporations will turn to the market. Japanese cost theorists however believe there is an intermediate area between market and hierarchy; this is the enterprise group. Their theory may be interesting, but they still have not found the principles of intercorporate relationships. These relationships are not intermediate; instead there is genuine principle in the intercorporate relation. The intercorporate relationship is not functional, but rather structural. The corporation is an early structure that was formed historically, and did not just appear out of nowhere to fulfil functional needs. Intercorporate relations are relational entities. Corporations conduct themselves functionally but they did not form to fulfil those exact functions. E. Cassirer tried to reduce substances to the functions but could not reduce all the substance to the function[2]. Also modern economic theory has tried to reduce substances (corporation) to functions (market mechanism), and so it too failed to find the principle of intercorporate relations.

Professor Imai tried to explain the Japanese enterprise group by a theory of information[3]. He said that the enterprise group is a kind of information club. But in my opinion the enterprise group is not an information club, in which member corporations come and go freely. Instead they have formed historically, so the structure is important.

Professor Nakatani wrote that enterprise groups were formed to share risks[4]. But the enterprise groups in Japan were formed after the 'zaibatsu' broke up. Their member corporations combined historically from the 'zaibatsu' era, and did not freely combine to share their risks. Their relationship was not formed freely or functionally in light of market mechanisms.

Professor R. Dore contrasts the long term relationships between corporations in Japan with short term relationships in the UK[5]. He said that these long term bonds provide one explanation for Japanese economic strength. He said long term relationships are characteristic of Japanese people. It was said that 'giri' (obligation) and 'ninjyo' (sympathy) is the tradition of the Japanese from the old ages. But the relationship among the individual people is different from the intercorporate relationships. We can make an analogy between the marriage and the merger or between the family and the parent-subsidiary company. But the principle is different. The merger of corporations is not a marriage because there is no feeling of love or hate between the corporations. One corporation sells or buys the other corporation, but the family cannot buy or sell children. Therefore there must be another principle in the intercorporate relation.

In the 1980s many people have attached great importance to the market mechanism not only in the USA, UK and Japan but also in the Soviet Union and China. In capitalist countries deregulation and privatization have become very popular. According to the theory of Hayek and Friedman they have prevailed. Even in the socialist countries the market mechanism has been regarded as important and necessary. But socialists do not find that intercorporate relationships are different from the market mechanism. They also do not lay much emphasis on the structure of the corporation (or enterprise). Professor R. Dore pinpointed 'obligate' or long term relationships as the reason why the Japanese economy is strong[6]. Some economists also considered the 'keiretsu' or the enterprise group in Japan. They believe that Japanese corporations put weight on quality control to maintain strength in long term relationships.

Perhaps the long term relationship is the reason why Japanese corporations are so strong and have succeeded. It may also be the reason that trade friction exists between Japan, the USA and the E.C. The US and the EC say that Japanese corporations prefer to buy from the related 'keiretsu' or enterprise group companies rather than from foreign companies when the price is low and the quality is better than Japanese products. Some Americans say that it is unfair that stabilization of stock ownership prevents the take-over while Japanese corporations have taken over many American companies. In the extensive and dense relationship among the Japanese corporations the foreign companies feel the difficulty. When the EC staff who wrote a report about the problem discussed with Japanese Foreign Ministry, the report was rejected as unreasonable. Many Japanese economists supported the rejection. But it is not reasonable, I think, they do not know the reality of the intercorporate relation in Japan.

Some economists say the situation is changing in Japan. It is true that the relationship between the banks and the corporations has undergone some change and that the ties between them are not as strong. Also many subcontract parts makers have begun to sell their products to the other parent companies. These changes, however have not become a general phenomenon. They are only partial. Some corporations have changed main banks but the main bank system has not changed. Some subcontractors have changed buyers, but they have not ceased to be subcontractors.

Intercorporate relationships in Japan have not changed significantly. Changing this situation, would require a second 'zaibatsu' dissolution. And this would require another General MacArthur.

Notes

[1] Akira Goto: Business Groups in a Market Economy. In *European Economic Review*, Vol 19

[2] Ernst Cassirer (1910) *Substanzbegriff und Funktionsbegriff* Verlagoon Bruno Cassirer

[3] Kenichi Imai (1986) Nihon no Kigyo network (in Japanese). In *Economics Today*, Autumn 1986

[4] Iwao Nakatani (1983) The Economic Role of Financial Corporate Grouping. In M Aoki ed (1983) *The Economic Analysis of the Japanese Firm*, Elsevier

[5] Ronald Dore (1987) *Taking Japan Seriously* Stanford University Press

[6] Op. cit.

Social Structure under the New Information Technology

Fumihiko Satofuka

The arrival of the so-called 'information society' has now been discussed at length for nearly a decade and a half. The coming of the 'information society' has been documented as a general tendency in both capitalist and socialist societies. This social change has also often been alluded to, under labels different from 'information society'.

Earlier it was designated the 'service society' (Marien 1983). Roughly the same connotation was used over a decade or so ago and called the Post Industrial Society (Bell 1974) or 'technocratic age' (Brzezinski 1970). The word 'electronic age', and the 'age of information', were labels given by Marshall McLuhan (1964) and 'knowledge society' was the label, given by Peter Drucker (1969).

The term 'information society' itself was first used in the late 1960's in Japan (Kohyama 1968), to be later used in a Japanese Government Plan, the Plan for an Information Society (Masuda 1981). Other variants included 'telematic society' (Martin 1982), as well as the 'Infoglut age' (Marien 1983). Although there is often an overlap and a conceptual unclarity in these formulations one can discern a common indication of a clear and dominant shift towards information related activities.

The movement towards automation generally and information based technology specifically, one should note is not a simple one. Just as the earlier industrial revolution was not spurred by one force such as profit alone, the present tendencies towards automation is influenced by a complex mixture of political, economic and intellectual influences (Noble 1986).

The trend towards an information society has occasionally been questioned and even termed a conspiracy (Roszak 1975). Roszack's view that there is a conspiracy of the Pentagon, vested interests, multinationals, and academics involved in the computer industry, to push the notion of an information society does not stand serious examination. Eastern European countries with a different social system are also feverishly adopting information related technologies and activities and theorising on them. Further, even 'counter culture' enthusiasts (Roszak wrote, the once considered definitive 'Making of a counter Culture' (Roszak 1975)) has had interesting debates on the nature of the information revolution such as in the counter-culture journal Under Currents (Garett and Wright 1979, Cambell 1979).

'Information' in some of these definitions does not necessarily mean computer based information, yet they include many clerical, technical, and service activities now rapidly encroached upon by the new technology. This shift to a predominant 'information sector', makes it easier for the speedy transition to information-machine based economic activities in these societies.

The dominance of the 'information sector' has arisen over the last hundred years. Porat (1978) has identified that between 1860 and 1906 the single largest group in the US was agricultural. Between 1906 and 1964 the industrial work force became predominant, reaching a peak of 40 percent in 1946. By the 1950's the proportion of the industrial work-force had begun to decline and by the 1970s it was only 25 percent. From 1954 to the present what he has described as information workers have become the largest group, a rise from 5 percent of the labour force in 1860 to nearly 50 percent by the mid-seventies.

Marc Porat did a major research project for the US Commerce Department on 'Information Economy' in 1975-76 and on the basis of this study he claimed that the US today is basically an information based economy and that by 1967 over one forth of the GNP was in the production, processing and distribution of information goods and services. Further he notes that over 21 percent of the GNP was in the production of information services for the internal use of public and private bureaucracies. By the beginning of the 1970s nearly half the US workers held jobs involved in the production, processing and distribution of symbols and were, hence, 'information workers'.

Porat described those industries that are engaged in the producing, processing and transmitting of either knowledge, communication and information as the 'primary information sector'. Among these industries are radio and television, newspapers, advertising, education, telecommunication, aspects of finance and insurance, libraries, and research and development firms, in the 'service department' of the information sector. In the 'goods department' he has included the hardware and business machines involved in this process.

Porat groups outside the 'primary information sector' those engaged in other aspects of information activity such as management, accounting, clerical work, and market information services. These constitute the 'secondary information sector', and both the primary and secondary sectors together amounted to nearly 50 percent of the GNP of the US. Clearly, a fundamental shift in the social structure was underway in developed countries.

Porat's methodology had been applied by the OECD (Organisation for Economic Co-operation and Development) to estimate the growth of the primary information sector in a number of countries as shown in the table below:

Percentage share of primary information sector in GDP at factor cost: Selected countries and dates

Australia	France	Japan	Sweden	UK	USA
(1968) 14.6	(1962) 16.0	(1960) 8.0		(1963) 16.0	(1958) 19.6
	(1972) 18.5	(1965) 14.4	(1970) 16.9	(1972) 22.0	(1967) 23.8
	(1974) 19.1	(1970) 18.8	(1975) 17.8		(1972) 24.8

It is clear that the economy has shifted towards information goods and services. A more disaggregated analysis of these trends indicate that almost 3/4 of this growth was due to information handling services (OECD 1981). A study of nine OECD member countries has also shown that there had been a substantial growth of the information labour force and that the information sector had become major sector for employment in all the countries surveyed.

Although the population profiles will not be identical of those engaged in different sectors in a country like the Soviet Union and in the United States - because of their different economic structure and stages of economic development -, in both cases there is generally a shift or more and more persons away from agriculture to energy intensive manufacturing, and subsequently to information handling activities. Indicative of the increasing information content in developed societies from a soviet perspective is the following prognosis from Dobrov, who looks at technology as a total system and notes that software and what he calls 'Orgware' (Organisational ware - 'the specialised organisational component of technology') would be dominant in the future. It is noteed that what he means by software and orgware are both information based systems.

There is no reason to doubt the logic of the shift that Porat has documented. The primary need of food, in today's technology is easily satisfied (even in developing countries if the available food is adequately distributed), and after the development of the basic industrial sector, the switch to information processing and handling seems natural. In such a system the major economic activity would be in the creation and exchange of information. This information and communication economy is being rapidly fuelled by developments in both communication and computer technology. Many of the activities in Porat's primary as well as secondary information sectors are being mechanised. The shift to an information society would thus be accompanied by a shift to electronic means of information processing. And, in fact, without such electronic means, a pervasive information society could not exist.

At the beginning of the early shifts towards information societies, the processing of information was largely done by humans with say, largely ink and paper. With the explosion of artificial information processing facilities, specially the expansion of computer and communication facilities, the processing of information has increasingly become electronic in character.

As yet there is no definite consensus about the emergent social structure corresponding to these technological and social changes, apart from the broad descriptions of the percentage of the population who will handle information.

The new information technologies, it is agreed, would have a major impact on 'private life, consumption, work places, organisations, institutional structures, political processes and societal development. There is hardly any social parameter which is not influenced by its introduction.' (Renn & Peters 1986). Current research has largely examined the impact of the new technology on employment and the quality of work, with slightly less research being done on its impact on the power structure, social relations and personal life (ibid).

Projections of the effect on employment vary widely. Thus, the emergence of computer technology over the last twenty five years has been seen as a three stage process. The first stage is the use of computers, word processors and copiers which expanded employment. The second is the merging of computers and telecommunication systems into 'telematic' systems, which also would increase employment. In the third phase, the telematic systems of separate institutions, became interlinked and develop organic links with each other which would then have a problematic impact on employment (Peitchinis 1983). It has also been hypothesised that the new technology would split the work into two camps of 'smart' jobs and 'dumb' jobs (Schwartz & Neikirk 1983). These and other structural projections indicate the problems of forecasting the future in detail and also the deep changers that are in store.

It appears that the introduction of the new information technology leads to the relative decline of the primary production sector in the national economy, and an increase in automation in the secondary and tertiary sectors and extensive computerisation of work and services. Both production activities as well as social interaction would be influenced by extensive communication networks and flexible work organisation, as well making possible flexible production techniques (ibid p.11). A discussion has also occurred about the possible democratising effects of the new technology, especially technology based on micro chips, which as well as replacing human labour, will permit a high degree of decentralisation of decision making. Because they are flexible and because they are run on software which is a socially changeable product, chip based technology would allow for a high degree of personal control (Garrett & Wright 1979).

Taking the opposite position, it has also been argued that instead of the democratic process becoming dominant through computers, the logic of market forces would relegate any democratic tendencies to a largely marginal position (Campbell 1979). Clearly, the detailed social effects of the computer and of the information revolution generally have not yet been worked out, so that a definite view on this is not possible.

However there are several studies which point to the deep changes in the economy, to the social stratification system and the attendant social roles. I will try to extract some of the salient findings in these studies, so that an overview can be given.

In the earlier section, I described how the stratification system in both capitalist and socialist industrial societies was, at a fundamental level, based on the answer to 'who controls the economy?'. In the case of capitalist countries, this control was held by an increasingly oligopolistic profit seeking private sector, whilst in the socialist countries it was by top level planners and the political elite. To both these control functions of the economy, at the level respectively of private profit seeking and macro planning - the new electronic and information revolution seems to be bringing great changes.

In developed market economies, formal systems of analysing profit maximisation are being increasingly used in investment decisions. A large number of technocrats who use formal systems of analysis are employed as financial advisors for many investment houses in the major centres of capital such as New York, Tokyo or London. More recently, the investment decisions about when to invest, under what conditions and so on, have been further formalised and incorporated in computer programmes. The consequence is that sometimes buying and selling at major stock exchanges, or at foreign exchange centres is at least partly done by automated computer programmes. At a signal from the economic environment, banks or computers buy or sell, shedding particular shares or currencies and acquiring others. This is a development of the last few years. It has been suggested that sudden recent (1987) changes in a number of financial markets have been due to such automatic computer programmes used for trading.

As far back as the 1960's, it had been realised that the vast number of economic decisions required in a complex centrally planned economy could no longer be left to unaided human effort. At one stage, it was calculated that the whole Soviet population would have to become planners if the system was to function efficiently. Hence, the recognition of the need for extensive computer use had been discussed for about 20 years (Lange (1973), Kornai (1965)). In fact as in classical Soviet planning exercises, mathematical models had always been used (Kantorowich 1959), the, extension to computerised modelling of the economy was but a natural growth.

Therefore clearly, in both capitalist and centrally planned economies, the new information technology is coming to play an increasing role at the macro level, replacing some of the information processing and decision making. At lower levels, computerised decision making is having an increasing impact on the various social strata of both capitalist and socialist countries. The computer is thus encroaching on the functions of macro economic planners, key corporate decision makers, various professionally and technically qualified workers, and various skilled and unskilled clerical and manual workers. The examples I give below of these encroachments are not meant=nt to be exhaustive.

Artificial Intelligence programmes existing for medical diagnosis for use by general practitioners or nurses would profoundly affect not only the role of the general practitioner or the nurse, but also that of the specialist physician whose function the AI programme would partly replace (Boden 1984). Similarly, legal expert programmes would affect the status of lawyers. The mystique of human expertise would thus be lessened. The potential for automating human creativity of the expert, could however in theory advance. Yet truly creative endeavours are undertaken only by few professional workers, most limiting themselves to routine, rule following work.

These new changes would also have an impact on the schooling system and instructional technology. With automated teaching devices, such subjects as reading, basic mathematics and descriptive botany and accounting for example, could effectively be done by technology (Dede 1981). The impact on the educational sector would be to alter their social roles and to shift the occupational structure.

The increased use of Computer Assisted Design (CAD) and Computer Assisted Manufacturing (CAM) would mean that current work being done today by engineers and architects would now be transferred to computer programmes. Here again it is possible that the extra leisure time, would leave these professionals extra potential time for 'creativity' (Hudson 1982).

With increased mechanisation of the office, there is a considerable degree of change in a wide variety of office work. Already the widespread use of desk top computer terminals and such devices as word processors is bringing deep changes. With the advent of devices as vary ideas voice activated typewriters to AI based information processing that will automatically retrieve, find and act on information, a wide variety of office jobs would be affected. These would cover the entire range of office work from the executive to the clerical level. Computers entered the banking sector a couple of decades ago and, electronic transfer of funds, and electronic dealing is already widespread in several financial centres. Here again there is a strong impact on existing social roles of clerical and other jobs, the existing functions and roles being usurped by the machine (Marti & Zeillenger 1982).

At the manufacturing level, the social role of the conventional blue collar worker is under threat. With the increasing use of robotics and the emergence of flexible manufacturing systems, a more-or-less complete integration of work is possible. A chain of devices from computer aided manufacturing to numerically controlled machines, would allow for the flexible manufacture of a wide variety of products (Johnson and Sasson 1986). As a result deep changes in the respective roles of humans and machines are occuring (Eayres & Miller 1982, Hudson 1982, Bylinsky 1983). Clearly, changes are currently underway affecting almost the totality of social roles in the classical industrial spectrum, whether it be in capitalist or socialist societies.

Our knowledge on the social impact of the new changes is still tentative. Yet, it is best to conclude this section by some considered remarks from a worker in the field, who describes the changes in the social roles which would probably occur in the developed countries within the next two decades at most, some probably within the next decade. A more strident advocate of Artificial Intelligence, say one committed to work on the fifth generation computer, would have anticipated greater changes.

'The economic impacts will be far reaching. Traditional manufacturing and clerical administrative jobs will be decimated. But new jobs will be created; some directly connected with new technology (like computer engineers and programmers), others made possible because people are freed to devote their time to services (caring professions, education, leisure). Whether there will be enough new jobs to compensate for the loss of old ones (as has always happened in the past, at least eventually) is however unclear, for AI can apply potentially to all jobs where personal human contact is not essential. New methods of work-sharing and income distribution will have to be worked out (with income not necessarily being closely linked to jobs). Radical structural changes in society are likely, and the transition phase will not be easy' (Boden 1985).

The emphasis in Boden's remarks is on current policy concerns such as unemployment. But implicit in her remarks - and the other evidence we have given earlier - is that hitherto social roles are being mechanised across the entire job spectrum. What was a human and social function, now becomes increasingly a machine function. The social roles in the occupational structure are now being partly replaced by machines and this process is increasing. A new 'occupational structure', a new set of roles based on machines seems to be hiving off from the human structure, although for quite some time traditional human skillls would still be required to tend to the machine stream (Johnston & Sasson 1986 pp. 84-85).

References

Bell, D (ed) (1967) *Toward the Year 2000: Work in Progress,* Beacon Press, Boston

Boden, M (1984) The Social Impact of Thinking Machines. In *Futures,* Febr.

Brzezinski, Z (1970) *Between Two Ages: America's Role in the Technotronic Era,* Viking Press, New York

Bylinsky, G & Moore, A H (1983) Flexible Manufacturing Systems. In *Fortune,* 21 Febr.

Campbell, A (1978) Hard Chips. In *Undercurrents,* No. 30, Oct.-Nov.

Dede, C (1981) Educational & Social Implications. In *Programmed Learning & Educational Technology,* Nov.

Drucker, P F (1969) *The Age of Discontinuity: Guidelines to our Changing Society,* Harper & Row, New York

Garrett, J & Wright, G (1978) Micro is Beautiful. In *Undercurrents,* No. 27

Hudson, C A (1982) Computers in Manufacturing. In *Science* 12 Febr.

Kantorovich, L V (1959) *The Best Use of Economic Resources,* Pergammon Press, Moscow

Kohyama, K (1968) *Introduction to Information Society Theory,* Winter, Chuo Koron

Kornai, J (1965) *Mathematical Programming as a Tool of Socialist Economic Planning.* Paper read at 1st World Congress of the Econometric Society, Sept., Rome

Lange, O (1967) The Computer and the Market. In Feinstein, C (ed) *Capitalism, Socialism and Economic Growth,* Cambridge University Press

Marien, M (1983) Some Questions for the Information Society. In *World Future Society Bulletin,* Sept.-Oct.

Marti, J & Zeilinger, A (1982) *Micros and Money: New Technology in Banking and Shopping*, Policy Studies Institure, London

Martin, J (1981) *Telematic Society: A Challange for Tomorow*, Prentice Hall, Englewood Cliffs, NJ.

Masuda, Y (1981) The Information Society as Post-Industrial Society. In *Society,* World Future Society, Bethesda

McLuhan, M (1964) *Understanding Media: The Extensions of Man*, McGraw Hill, New York

Miller, J G (1977) *Living Systems*, McGraw Hill, New York

Noble, D F (1986) *Forces of Production: A Social History of Industrial Automation,* Oxford University Press, New York

Peitchinis, S G (1983) *Computer Technology & Employment*, St. Martin's Press, New York

Porat, M (1978) *Introduction to Information Economics,* USA

Renn, O & Peters, H P (1986) *Micro and Macrosociological Consequences of New Informations Technologies*. Paper read at XIth World Congress of Sociology, Aug., New Dehli

Roszak, T (1975) *The Making of a Counter Culture,* Faber, London

Schwartz, G F & Neikirk, W (1983) *The Work Revolution*, Rawson Ass., New York

Can the Family Survive the New Communications Media? 'Human Centredness' and the Challenge of Consumer Technologies

David Smith

Scenario ...

... which could be set in Tokyo, San Francisco, London, Rio, Delhi, or anywhere else that the international technoculture has taken root ...

It is Saturday evening and the Watanabe family are at home. Father is using his lap-top computer to prepare a sales report for the Bucho of his electronics firm. (He is lucky to be home at all: often his work takes him away for weeks on end, or keeps him occupied until late at night). Mother is watching a video whilst preparing the evening meal. Teenage son is playing a computer game in his room. Ten-year-old daughter is watching a cartoon programme on her own television set. There is very little conversation: the family is at home but not together.

Introduction

The impact of the conventional media on family relationships has been a matter of concern for many years. But the penetration of the new information and communications technologies (NICTs) into the home is an order of magnitude removed from earlier developments. The NICTs undoubtedly offer many exciting possibilities for new entertainment media, as well as opportunities for changing the pattern of working life, but they may also be imposing new strains on family life in particular and on society in general.

This paper will examine some of the stresses which the increasing convergence of computers and communications media may be placing on the family, and thence on society. It will argue that 'human-centredness' goes far beyond manufacturing technology, and it will explore some of the ways in which this principle might be exploited to design technological environments which could counterbalance social fragmentation.

Technology and the Family: A Love - Hate Relationship?

There is a long tradition of concern about the impact of technical innovations on family life and relationships. Much of this concern has, arguably, reflected male responses to the gradual emancipation of women from a daily round of tasks and responsibilities which effectively tied them to hearth and home: it is surely no coincidence that the washing machine, the vacuum cleaner and convenience foods have all been condemned as potential catalysts of moral decline through their supposed encouragement of (female!) laziness.

Disturbing though the instant soba may have been, it is technologies of mass communication and entertainment which have been most frequently and clearly identified as the ultimate threats to civilisation. This is nothing new. The Greek philosopher Plato was all for the total censorship of art, music, literature and drama, whilst in nineteenth century Britain (and elsewhere), just about every social ill from alcoholism to street crime was blamed on such early commercial exploitations of growing mass print literacy as the 'penny dreadful' comic paper. There is even some evidence[1] that English magistrates in the last century were particularly lenient to criminals who confessed to having been led astray by their choice of reading matter!

But it was the penetration of mass entertainment into the home which initiated a debate which is still vigorously contested. The various malign influences which had been attributed to the early cinema resurfaced with a vengeance a few years later when radio came on the scene. Concerns ranged from worries about the demise of the 'art of conversation' to the political 'corruption' of the working classes. Predictions about the adverse effects of broadcast television were inevitably little short of apocalyptic.

The advent of the computer has added yet another villain to the story. The stereotypic 'couch potato', mesmerised by the television screen into something close to total brain death, is now joined in popular imagination by the computer freak, zapping aliens far into the night. But it is still television that is most widely seen as the great contemporary corruptor of youth, to the extent that an enormous sustained (and largely fruitless) research effort has been directed towards trying to demonstrate its demonic powers.

Despite all this, the fact remains that, however much we may distrust the products of advanced technology, we aren't willing to live without them if we can possibly avoid doing so. We might feel that dishwashers somehow weaken our moral fibre, but that doesn't prevent us buying them if we can possibly afford it. We may see computers as faintly sinister devices, but we rush to buy home micros. We may suspect that videos are harmful to us and our children, but we still buy and rent them in ever increasing numbers. We may resent every minute of it, but between the ages of two and sixty five, many of us will still spend nine years in front of a television...

> *'... much as we try to love them, we know that we can't trust them. They're aliens which have requisitioned our houses, and we'd all like to fight back by damaging them - if only we didn't own them ...'(Conrad, 1982[2])*

Research over the past ten years or so has indicated the complexity of our attitudes to 'domestic technology' and of the social interactions which take place around it. The ambivalent 'love-hate' attitude described above is simply one aspect of a more general phenomenon which we will have to take very seriously if we are ever going to understand the influence of technology on the evolution of the family - and vice-versa.

One thing seems true, however. There has been a great deal of very simplistic analysis, and communications media have been blamed for many perceived social ills on the basis of 'post hoc, propter hoc'. For example, the fragmentation of the family referred to in my 'scenario' could easily be seen as a simple direct consequence of the new media per se. But in fact it reflects, among other major factors, a process of well-targetted market segmentation, which is essential to the efficient operation of commerce and industry, and which would not work if the social structures within which it operates were not already there - at least in embryo.

Whether or not we like all the implications of this process, we have to accept that it is a trend which will inevitably intensify as the possibilities for new generations of interactive wide-band communications systems come to be exploited in the market place. Despite our distrust, despite our obvious reservations, we are unlikely to refuse to buy devices which will soon make it possible, for example, to tune into a TV channel and then to select one of eight parallel audio tracks. Given that we can afford to buy the devices in the first place, we seem perfectly content to accept the social logic of market segmentation, including the process of segmentation which classifies us according to our consumption or non-consumption of devices!

Whatever role current mass media may be playing in weakening the cohesion of the family unit, and that is a matter for debate, they represent only part of a complex series of technological, social and political factors. Western radio and television broadcasting, for example, have developed in the context of an already atomised society, each fragmented family unit living in its private home. The British educational sociologist Frank Musgrove observed nearly a quarter of a century ago[3]:

> *'The 'home-centred society' was not unheralded... it is not simply a by-product of television... modern technics have aided a long-established social trend. The motor car is a detachable parlour which enables the family to move off to the moors, the city or the coast still effectively insulated from all outside human contact.'*

There is a limit to the power and effectiveness of social engineering, and in any case, social relationships are far too complex for minor tinkering to have much effect (whilst the recent history of Eastern Europe seems to suggest the futility of state direction):

> *'...there is an interdependence of technological advance on the one hand and social change on the other that is sometimes of a very subtle and complex, yet far-reaching nature, and that is frequently quite unplanned...' (Schaffer, 1985[4])*

It would be a serious mistake, therefore, to assume that we can 'preserve' family life by regulation or legislation, or that there might be a simple technological 'fix' which could somehow switch social evolution onto a more desirable track.

'Technology and Society': One View

Yukio Mishima once observed that

> *'... advice is free. We may be reluctant to lend a hundred Yen, but we are no more reluctant to part with a bit of free advice than with water ...'*

We are all aware that we no longer live in a static world. I'm sure that the Japanese people need no reminders of the culture shock associated with the arrival of the Black Ships, but it is very important to emphasise that the road to the present situation where so much of Japanese industry has succeeded in establishing itself as 'Ichiban' has been neither easy nor inevitable.

We seem to owe the much-misused phrase 'The Information Society' to a study by Professor Yoneji Masuda, presented to the Japanese Government in 1972[5]: before the advent of the micro-computer. Most early thinking, and we might point to the importance of Daniel Bell's seminal analysis[6] here, concentrated on the economic and commercial changes which were to be expected in consequence of coming generations of computing and other data processing machines. The most significant of these anticipated changes related to the refocussing of economic activity in advanced countries from manufacturing to the so-called 'information industries' and with the emergence of information as the basis for a 'quinary' sector of the economy.

It has become almost conventional to see changes in society in terms of the second- or third-order effects of patterns of economic activity (with all the implied difficulties of prediction), rather than in terms of human relationships per se. Much attention has been given to negative social factors such as the mass impact of structural unemployment and technological displacement and to positive factors such as the decline of hard, dangerous and dirty occupations and a general increase in 'leisure'. This level of description reflects an overall concern with aggregate society, rather than with the lives of citizens, though it is clearly recognised that various factors must inevitably affect individual people. But as Mike Cooley's[7] trail-blazing work suggests, it is not enough to deal with people as abstract social aggregates or idealised subcultures. We also need to understand the new technologies from the points of view of individuals and their families.

Professor Takao Nuki of Musashi University has recently put forward the image of technology as providing the 'engine' and culture as providing the 'control systems' in our daily lives. He differentiates clearly between 'devices' and 'technology', seeing the latter as a series of intellectual factors (knowledge, attitudes etc) which condition and enable our use and control of devices. This is a valuable point of view, stressing as it does the roles of both social and personal cultural outlooks in shaping our response to new technological developments.

There are many situations in contemporary society where our access to devices outstrips our ability to understand, and hence to control them: indeed, many contemporary developments in 'social technology' can be compared to putting a Formula One engine in a go-kart! One of the major tenets of 'Human Centredness' is that the evolution of technology (however defined) should be determined by human needs, rather than vice-versa. If we neglect this principle we run the risk of diverting attention away from people and towards organisational and managerial aspects of change. We can end up discussing control when we do not understand what we are controlling.

Fast Forward!, Fast Forward!

Last year my local cinema was reshowing one of the 'Superman' films to a largely teenage audience. The story had just reached the 'romantic interest' point. Superman was kissing Lois Lane! But instead of simply either waiting impatiently for the action to be resumed or else wallowing in the pleasures of vicarious lust, the audience started chanting 'Fast Forward!, Fast Forward!, Fast Forward!' The young people in the audience were both 'film-literate' and sophisticated enough to derive an appropriate metaphor from their familiarity with video-recording technology. But there was something else, too: something which is important in the context of this paper.

The audience were not merely receiving a text in the passive sense understood by many critics of current media. There was an implicit expectation of control, in which they were sharing a perception of themselves as having a potentially interactive relationship with technology in a cultural sense rather than simply in terms of 'human-machine interaction'. We can expect this perception to grow dramatically as Integrated Wideband Communications Systems penetrate homes and workplaces, and it is vital that we should not allow our vision to be constricted by the limitations of current examples of supposedly 'interactive' systems.

The new Information and Communications Technologies (NICTs) may influence interpersonal communication by shifting perceptions of both what it is possible to say and how it may be said. Traditional societies have always transmitted fundamental cultural information in forms such as dance, music, myths, drama and iconic symbolism, and we could witness a rediscovery of these modes of communication as everyday forms, rather than high 'arts'. Young people today are, as it were, growing up alongside the new interactive media. It is entirely probable that they are unconsciously participating in the creation of new symbolic languages concerned with the sequencing and association of images, sounds and conventional text. Multimedia technologies will offer them the capability for expression in a complex fusion of verbal and non-verbal forms, and we can only guess what the social outcomes of this process might eventually be. We are certainly not justified in nipping the whole thing in the bud in an attempt to retain control of future society. We are, however, entitled to participate democratically in the process of 'shaping' the new technology to mutually agreed purposes.

Technology and Society: A Slightly Different View

UK sociotechnical specialist Ian Miles has recently[8] drawn attention to the fact that private households have gradually become more and more capital-intensive over the past few decades. Miles identifies two broad classes of relevant innovation: goods based on cheap energy and motive power and goods based on electronics for entertainment, news and communications, pointing not only to the growing complexity of many household devices but also to their increasing convergence of function. He argues that much home technology represents a more or less direct transfer from industrial or military applications, and calls attention to the parallels which he sees between applications of IT to the informational aspects of industrial work and its application to comparable functions in the household.

This analogy between the 'technologisation' of the home and that of the workplace is an interesting one: the more so as the home actually becomes the workplace for an increasing number of people. It suggests that families may be using home technology in ways which derive directly or otherwise from industrial practices, rather than from social cultural values. Thus, washing machines are generally seen as liberating women from routine drudgery: and so they do. But they may also, in fact, bring with them implicit needs for productivity and efficient use of deployed capital (reinforced by advertising) which pressurise housewives into washing things far more intensively than is strictly necessary either for appearance or hygiene (resulting in erosion of the 'leisure' apparently gained by buying the hardware in the first place and, incidentally, in accelerated deterioration of many natural fabrics).

But this should not be seen as a one-way process. The converse is also likely. It is highly probable that patterns of thinking and acting developed in interaction with domestic technologies will carry over into the world of work. One example can be found in gender-related attitudes. There is now a considerable body of evidence[9] (at least in Europe and the USA) which points to clear differentiation ('gender colouring') between the aspects of domestic technologies over which male and female family members are likely to establish effective. The indications are that men are more likely than women to get beyond simple operational control of a wide range of devices. This perceived relationship could be one of the many factors contributing to the unwillingness of young women to take up careers in science and technology in countries such as the UK; though we can perhaps hope that its impact will decline in the future.

Another (and possibly even more significant) effect on industry could derive from changes in deference relationships within the family, catalysed in part by NICTs.

The age-old confucian ideal of 'three generations under one roof' is fast becoming a thing of the past for a whole variety of reasons including, in the USA, the lure of the Florida coast! An increasing percentage of young people are growing up in families where the traditional pattern of deference towards older family members is not strongly reinforced. Even in Japan, there is a shift in attitudes towards 'ie-tsuki, kah-tsuki, baba-nuki!" There is, consequently, a shift in the socialisation of the child away from the influence of a seniority hierarchy within and around the immediate family, towards increased salience of the peer group on one hand and towards an almost asocial individualism on the other.

Ironically, the good old fashioned one-per-home television set can help to counteract this change by introducing a focus for conflict (when to turn on, which channel to watch etc.) which is usually resolved according to age and status criteria. The availability of 'private' television sets and other devices defuses some of the minor challenges to authority which are part of normal family life. Instead of serving as a totem of family authority, ritually controlled by the head of the household, the remote control simply becomes yet another private tool for daily living.

Children may also be deriving a rather jaundiced view of adults and adult society from the media:

'... because they are over-exposed on television to literally hundreds of adults engaged in all kinds of action, children may be losing any sense of awe or of mystique or special quality which they might have enjoyed when children came across only a few adults. In consequence, adults may no longer be perceived, even by very young children, as authority figures ...' (Steed & Lawrence, 1986[10])

The effect of all this is experienced in the home, where young children may be the only family members fully competent in handling electronic devices, in education, where the authority of the teacher is rapidly being eroded, and in the world of work, where management structures based on idealised clan/family deference hierarchies (such as corporate paternalism) are increasingly inoperable.

Nevertheless, there is good reason to hope that the ground has been prepared for the germination of a countervailing tendency to the vision of totalitarian control which many observers perceive in:

'... a market-dominated development that broke up the community and sealed individuals into isolated physical and physiological cubicles. Transmission from centralised message centres to atomised individual receivers serves and replicates the one-way flow that is built into the system and that separates the rulers from the ruled.' (Schiller, 1976[11], pp 51-52)

In fact, some writers[12] see this as a failure of the market, and argue that NICTs will eventually enable consumers to exercise true freedom of choice in an efficient market.

We are, as Professor Nuki rightly sees, ultimately talking about culture. This point is also well appreciated by those organisations such as the Swedish Working Life Centre (Arbetslivscentrum), which has given a great deal of serious attention to the issues of working life culture.

'Conviviality' and Human Centredness

One really important issue here is whether we see ourselves as controlling our technology or being controlled by it, and this is very far from being a new question. The critical significance of the intrinsic nature of our tools for both society and individuals was argued by the radical educator Ivan Illich:

'Tools are intrinsic to social relationships. An individual relates himself in action to his society through the use of tools that he actively masters or by which he is passively acted upon. To the degree that he masters his tools, he can invest the world with his meaning; to the degree that he is mastered by his tools, the shape of the tool determines his own self-image. Convivial tools are those which give each person who uses them the greatest opportunity to invest the environment with the fruits of his or her vision'. (Ivan Illich, 1973[13] p. 21)

Margaret Boden[14] made a really rather similar point:

> *'... if a man's self-image represents himself to himself as an autonomous purposive creature capable of pursuing certain ends, then it can ... generate choices and guide his action accordingly. If depersonalisation of self-image occurs, so that the self is no longer seen as a truly purposive system, then relatively inhuman pathological behaviour can be expected.'*

The problem is, how do we design devices and technological systems to be 'convivial', in Illich's sense, or to avoid the psychological damage discussed by Boden and others without falling into the trap of attempting wholesale social engineering? I believe that our future prosperity, personal happiness and social cohesion rest to a significant extent on our finding the right answers to these questions.

The obstacles here are not primarily technical:

> *'We have proved that it is technically feasible to do many exotic and previously unheard-of things ... While there are still real technical barriers... we have learned that the more difficult questions involve assessing social needs and demands, how to finance services, differential access to services, 'institutionalisation', training and so on. These are human, social and political questions, not primarily* based on technical barriers ...' *Ladislav Cerych (1985)*[15]

The barriers are, rather, technological, in the broader cultural sense. This is just where the movement for 'Human-Centredness' is beginning to achieve some major attitude changes. If we cannot engineer social change, we can at least work to offer devices and ideas which support or enable our conceptions of how things ought to be. We can work at two levels, that of education and awareness, and at the level of practical systems design.

In terms of general awareness, the message is beginning to get across. More and more people are beginning to understand that technology is not uncontrollable except by a privileged elite, that it is one social process among many, and that society and technology are mutually interactive. The 'Green' movement in Europe is one manifestation of this growing awareness, but it is one which carries a risk of rejecting the benefits of advanced technologies in favour of a curious 'lost paradise' utopianism unless there are clear demonstrations of how NICTs fit in. The work of Achille Ardigò[16] and his colleagues, with their conception of 'social citizenship' as supported by an appropriate technological infrastructure is extremely important in this respect.

Education as a state institution, on the other hand, is very difficult indeed to change, and has proved remarkably unresponsive to the epochal changes associated with the advent of the 'Information Society'. In one respect, this is just as it should be. For more than a century, formal education has been one of our most crucial mechanisms for cultural knowledge transfer, and has therefore needed (and still needs!) a certain resilience to intellectual and political fashion. But many aspects of formal educational systems are now increasingly out of touch: with the realities of contemporary social life; with the specific requirements of effective participation in all aspects of citizenship; and not least, with the possibilities for positive development inherent in the NICTs.

This all points to the increasing instability that Pogrow (1983)[17] labelled 'environmental collapse'. Using technology in a 'Human Centred' way to revitalise education (supposedly an inherently human centred social process!), perhaps by refocussing it around its once traditional family context, will be one of our greatest challenges in the coming decade. Karamjit Gill has pointed the way here with the concepts of 'participatory design' underlying his Parosi project[18] and the subsequent work of the SEAKE Centre.

At the corporate level, there is a clear need to demonstrate the validity of the principle by designing, developing and implementing practical Human Centred systems. Mike Cooley, Harold Rosenbrock and Felix Rauner among others in Europe have played a crucial role in advancing the case for human centred design by creating innovative 'anthropocentric' machine tools for manufacturing industry. But the far greater challenge facing us is the extension of the ideal of human centredness into the design of 'everyday' devices and working or leisure environments.

We can either design technical systems and devices which are in sympathy with human needs, aspirations and social interactions, or we can effectively destabilise the fundamental interpersonal processes on which societal and corporate coherence and effectiveness ultimately depend. This is where we need to be specially sensitive to the complex interaction between family life and industrial culture.

Conclusion

Despite much gloom and foreboding, 'the family' is not so seriously threatened as is often believed. That is not to say that there are grounds for complacency or any need to relax our vigilance, but all the indications are that the basic unit of social organisation is evolving fast in the face of extraordinary pressures and intriguing new opportunities.

Musgrove[19] has even seen the modern family in quite a different light:

> *'... So successful is the modern conjugal or nuclear family that it is a threat to society. It has lost its historic sociability. It is no longer a link between diverse social groups. It has imprisoned the father, a traditional liaison officer with the wider society outside the family, and it has imprisoned the children, even using the school as an extension of itself ...'*

The NICTs seem to offer ways of breaking out of this prison, but at the cost of introducing new change-agents into an already complex and increasingly unstable social system. Changing patterns of family organisation and home life can have far-reaching effects on the fundamental attitudes which enable the management of high technology industries to function at all. Perhaps the real issue we have to face is not *can the family survive technology?*, but rather *can technology survive the family?* !

The problem before us is therefore to adapt our social and commercial infrastructures to reflect changes in the ways in which individual citizens view their relationships with each other and with the tools which technology puts at their disposal. In order to achieve this, we need new principles of systems design and implementation. The ideal of 'Human Centredness' is a major step forward in this respect.

It means not trying to censor the new media, or applying 'big brother' constraints on their development and application, but instead asking how we can use the enormous potential power of integrated broad-band communications systems to enable, support and enhance the widest possible involvement of people of all ages and conditions in the workings of society. Ardigò's idea that NICTs could comprise a new branch of public and private social welfare deserves thorough exploration in this context.

It used to be said that 'The family that prays together stays together'. Maybe that's true and maybe it isn't, but in an era when prayer seems rather less than wildly popular, we can hope that a shared active family participation in citizenship will provide a new focus for social evolution, perhaps by supporting new modes of social communication[20].

Afterthought

Cicada - did it
chirp till it
knew nothing else?

(Basho)

Notes

[1] Pearson G (1983) *Hooligan: a history of respectable fears* London, MacMillan.

[2] Conrad P (1982) *Television: the medium and its manners* London, Routledge & Kegan Paul.

[3] Musgrove F (1966) *The family, education and society* London, Routledge & Kegan Paul.

[4] Schaffer R (1985) *Growing up in a technological society* Universities Quarterly. Winter 1985/86 31-34.

[5] Masuda Y (1972) *The Information Society as Post-Industrial Society* (reprinted) World Future Society, Bethesda, 1981.

[6] Bell D (1980: reprinted) The social framework of the information society. In Forrester T (ed) *The Information Technology Revolution* Oxford, Blackwell, pp 500-549

[7] Cooley M (1987) *Architect or bee?* London, Hogarth

[8] Miles I (1988) *Home Informatics: information technology and the transformation of everyday life* London, Frances Pinter.

[9] See, for example, Gray A (1986) *Video recorders in the home: women's work and boys' toys* Paper given at International Television Studies Conference, London. See also,

Morley D (1986) *Family Television: cultural power and domestic leisure* London, Comedia

[10] Steed D & Lawrence J (1986) Why video nasties are bad for teachers. In *Education*. 18 April 1986.

[11] Schiller H (1976) *Communication and Cultural Domination* White Plains New York, International Arts & Science Press.

[12] Malone TW, Yates J & Benjamin R (1987) Electronic markets and electronic hierarchies. In *Communications of the ACM* 30 (6) 484-497.

[13] Illich I (1973) *Tools for Conviviality* London, Calder & Boyars.

[14] Boden M (1977) Social Implications of Intelligent Machines. In *Radio & Electronic Engineer* 47

[15] Cerych L (1985) Problems arising from the use of new technologies in education. In *European Journal of Education* 20 (2/3) 223-232.

[16] Ardigo A (1989) *New Technology and Social Citizenship*. Paper given at International Workshop on Human Centred Systems Design, Brighton, 1989.

[17] Pogrow S (1983) *Education in the computer age: issues of policy, practise and reform* Beverley Hills, Sage.

[18] Gill KS (1990) Cultures, Language, Mediation. In Goranzon B & Florin M (eds) *Artificial Intelligence, Culture and Language: On Education and Work* London, Springer-Verlag. pp 171-183.

[19] op. cit. p.55

[20] Mazzoli G (1989) *A New Communication in a New Social System* Paper given at International Workshop on Human Centred Systems Design, Brighton, 1989.

Anthropocentric Production Systems in the Context of European Integration

Werner Wobbe

Introduction

The Commission of the European Community runs a research programme, *FAST*, (Forecasting and Assessment in Science and Technology) to establish priorities and orientations for EC science and technology policy.

The second FAST programme ended in 1988/89 and proposed a European Community R&D programme on 'Human Work in Advanced Technical Environments' in order to develop anthropocentric technologies. This proposal intended to broaden the scope of Community actions beyond its emphasis on IT products and modernising old industries. The principle message of FAST was that the tryadic competition between USA, Japan and Europe, concentrating on technology development itself, is not sufficient for the welfare of Europe. Cultural differences in the European member countries support different implementation modes for specific technologies, increasing imbalances within the community instead of making it more coherent. Furthermore it was called into question whether the innovation and application modes of traditional technologies are as efficient as claimed. Therefore better adapted technologies, taking into account the skills of users and new organisation modes which change the traditional tayloristic factory is thought a better way to cope with future developments. This is a typical FAST approach, namely, exploring socio-economic developments first, and reasoning afterwards about science and technology policy.

In the following the FAST programme in general is presented as well as some research results.

FAST Objectives

The FAST programme, launched by the European Communities in 1979, is a tool for developing long-term orientation for R&D at community level and contributes to the definition of new strategic priorities for Community policies associated with scientific and technological developments. FAST has two main tasks:

- ***a research task:** the analysis of scientific and technical changes in their many dimensions - economic, social and political.*

- *an organisational task: the strengthening of the basis of current European forecasting and assessment thinking by encouraging the creation of ad hoc cooperative networks between researchers, civil servants, industrialists, and concerned social groups; on a European scale.*

The main orientation of the FAST approach is towards the social and individual changes which are being brought about by the application of the new knowledge and technology. A technology-driven analytical approach (technological development first, then afterwards consideration of its impact on the economy and society) is not the most pertinent and useful approach to serve the interests of policy-makers, who expect to derive from technology assessment and forecasting a better understanding of ongoing and potential complex social changes and clear indications for options and strategic choices.

The FAST forecasting activities since 1989 are a reoriented follow-up to the previous FAST programmes. It has been allocated 11mio$ for the research activities, personnel, scientific fellows, conferences, research networks, etc. The aim is to provide the Commission with global analyses and long term projections in relation to the Community's major objectives for the 1990s, namely the creation of a single European market and strengthening economic and social cohesion within the community - seen in the light of world wide economic and social developments.

Conclusions Drawn from FAST Research[1]

More than two hundred studies have been carried out in the second cycle of the FAST programme which have contributed to R&D orientations established by FAST. Concerning the industrial future, FAST studies point out that employment problems and changing employment conditions are caused less by technological developments than by *structural changes in the reorganisation of our industrial society* referred to as *meta-industrialisation.* FAST does not therefore tend to favour purely re-industrialisation which policy makers have supported over the past few years, but recommends that strategic alliances between producer services and industries and their suppliers be strengthened.

The fully automated workerless factory is not expected to materialise, but new forms of divisions of work and cooperation will result through *computer integrated manufacturing.* It is important that the potential of information technologies is used in this context to adapt to the abilities of its users.

To stabilise Europe's technological position in the world, it is recommend that, besides supporting the information and communication technologies in the ongoing Community programmes, more attention is paid to new materials and light technologies.

In addition to EC scientific and technological objectives, the safeguarding of European manufacturing knowledge and its human resources has been given a priority position, by establishing R&D programmes.

R&D for industrial application should in future pay much more attention to human aspects. Technologies should be designed such as to assist man to achieve his practical ends, rather than treating him as a servant of technology. Such an approach demands the development of *computer and human integrated systems* (CHIS), and promises to be a much more fruitful approach in generating productive, robust and controllable technologies than a purely technical approach.

Industrial restructuring has to be faced and adaptation measures for innovation have to be developed as follows. Traditional industrial structures are changing with regard to the principles of traditional mass production, notably a high degree of division of labour, dedicated automation, product life cycles, large production facilities, hierarchical enterprise organisation. The policy of economy of scale is in transition towards economy of scope or variety. This has serious implications for enterprise organisation, technology, human resource issues, and industrial cooperation schemes, and it leads to *new forms of industrial structures.*

One element of that new industrial structure is *meta-industrial cooperation.* It means that core enterprises are operating with a periphery of subcontractors for goods and industrial services. This presents changes which could create serious problems, particularly for small firms. Business services are of increasing importance. Telematic facilities are a strategic means of aiding this process, as well as the cooperation between small enterprises which will depend on information access similar to that of large firms, and on long term relationships as well as on a publicly provided infrastructure.

New forms of manufacturing - the *flexible specialisation pattern* - will continue to increase. These developments increasingly favour small production units. In terms of world competition, they seem to be becoming a dominant manufacturing pattern for Europe, able to adapt to quickly changing markets and delivering complex and quality advanced products. These production units are based on programmable technologies, skilled workers and intelligent organisation and knowledge based management. Key elements are skills, factory and work organisation. To aid this process in Europe, anthropocentric technologies, computer and human-integrated manufacturing (CHIM), flexible forms of organisation - *orgware and orgknow* - as well as training measures have to be developed.

Human Work in Advanced Technical Environments

In consequence of this orientation pointers FAST proposed last year a R&D programme for anthropocentric technologies for the community. A working group lead by Mike Cooley[2] was charged to give a first outline which has been further developed by FAST[3]. The particular reasoning behind the programme initiative is as follows:

Productivity, economic success and social welfare in the beginning of the 21st Century will be dependent on a bundle of strategic factors in which technology has carefully to be embedded. Already by now, flexibility, changing markets, the role of SME'S, environmental concern, the use of new materials, education, ageing of society and the role of young people, mis-investment, vulnerability of 'big' technical systems, have led enterprises and public bodies to reconsider the question of scientific and technological development, application and diffusion.

It has become evident that productivity and competitiveness will increasingly depend on the link between the skills and ingenuity of humans and advanced and complex forms of technology.

New R&D programmes have therefore to be based on the 'valorisation' of human factors in production environments by designing, developing and applying technologies which can be called *anthropocentric technologies and systems.*

The key focus of an R&D programme would be:

- *how to optimally achieve higher productivity and competitiveness in manufacturing and*
- *how to avoid misguided investment.*

The programme would deal with an issue which is found on the agenda of advanced enterprises today. Experiences with complex information technology based systems have not led to the results expected. 20-50% of investments have not reached their aims and are often taken out of service.

Such R&D action will bridge an important research gap, i.e. fundamental and applied research devoted to the identification and definition of the scientific, technical and organisational prerequisites to designing and developing highly advanced technical production systems which conform with human, social and environmental requirements.

The subject matter of the research is a new generation of *anthropocentric technologies and systems* which will enhance human skills and competence. While being conceived and designed in Europe to respond to the variety of social structures and cultures within European societies, this proposal is also open to cooperation with non-community research groups particularly from the less developed countries.

Specific Reasons Demanding a New Approach

Several socio-economic and technical reasons support of the development of anthropocentric technologies and systems.

The socio-economic reasons reflect the particular nature of the European production structure, and are related to problems posed by SME'S, shifting markets, changing demands, broader product variety, environmental factors, education and demographic development. Technological reasons are concerned with flexibility, vulnerability and misguided investments.

Socio-economic Reasons

SME's

Contrary to the situation in the United States and Japan, much of European industry is characterized by SME's. Indeed, in some Member States the entire industry is composed of SME's. In general, SME's provide around 75% of total employment and 65% of GNP in the community. Publicly funded production technologies have too often started from organisational structures and production modes of large firms and reduced in scale afterwards to be adopted by SME's. The holistic organisation and the skill base of small production units have to be considered as a starting point for technological development and design.

Markets

Demand for industrial consumer goods is shifting from steady expansion to stagnation. The growing wealth in industrialised countries has led to higher prices, and more individualised and customized products. The European industrial core might, therefore, be better off to specialise in those areas which require order-bound manufacturing and hence experience in flexible specialisation based on the flexible production systems and skilled workforce at their disposal. Long production runs and mass production have given way to forms of production with a broad scope of variety and with high quality features.

Environmental and Material Problems

The 1990s will be marked by a growing demand for processes which involve the lowest possible release of eco-impacting materials to the environment and the highest possible degree of recyclability, as well as use of materials together with the highest achievable energy efficiencies. This implies not merely new forms of materials, but new forms of production and new product ranges which facilitate repair, recycling and refurbishment. There are grounds for believing that anthropocentric technologies will actually facilitate production processes and product ranges which display more energy and ecologically desirable characteristics.

Demographic Development

According to UN projections the population of Europe over the age of 60 will increase by 40% between 1989 and 2000. The traditional development of technology rejects existing skills and creates a demand for skills that do not exist. This process of skill transformation in industries has serious implications in connection with the demographic change.

Education

The tendency for education to concentrate on narrow, specialist areas is counterproductive and must give way to more holistic forms. The concern should be education rather than training. Above all, education should be the transmission of a culture which values proactive, sensitive, creative human beings. The education of engineers and managers must equip them to design systems and produce organisational forms which are skill-enhancing.

Technological Reasons

The concept of *Total Automation* runs into several difficulties making its long-term success doubtful:

Flexibility

Firms following the strategy of 'total automation' would suffer from relative inflexibility with respect to alterations of batches and process innovation. The more complex automation systems become, the more inflexible (including maintenance and break-down) they seem to grow.

Vulnerability

Indiscriminate attempts to replace humans by machines have led to serious difficulties with machine dependent systems which are vulnerable to disturbances and which frequently lack robustness and flexibility. Disruptions, large scale accidents, systems breakdowns, low capacity usage are different forms of unexpected vulnerability. They are mainly caused by inherent problems in the design of complex systems, deterioration from norms, and ignorance of human communication systems.

Misled Investment

World wide investment in Flexible Automation and CIM is considerable (in 1977, 35 billion was invested in automation technologies and this sum is expected to increase to 77 billion in 1991). However economic problem solving applications seem not to lie in bigger and more complex systems, but in new organisational concepts, and smarter technologies calling for a new link between the skills and ingenuity of humans and advanced forms of technology.

Are All Cultures Receptive to Anthropocentric Technologies?

Although the advantages of Anthropocentric Technologies seem to be plausible it has to be recognised that technology design and its implementation are heavily cultural biased. We distinguish for analytical purposes at least two cultures: the *technocentric* and the *anthropocentric*. These cultures could be characterised schematically as follows:

THE SOCIETAL TECHNOCENTRIC PATHWAY :	THE ANTHROPOCENTRIC WAY IN SOCIETY :
o WHITE COLOR DOMINANCE IN MANUFACTURING	o BROADLY EDUCATED HUMAN CAPITAL AT ALL LEVELS
o THE BELIEF ONLY ENGINEERS COULD SOLVE MANUFACTURING PROBLEMS	o ACCEPTANCE OF BLUE COLOR EXPERTISE
o PRODUCTIVITY RAISES BY TECHNOLOGY	o PRODUCTIVITY ACHIEVEMENTS BY ORGANISATION
o LOW BLUE COLOR TRAINING - HIGH WHITE COLOR EDUCATION	o DISPERSED MARKETS WITH DIFFERENT CULTURAL FLAVOURS
o HEAVY INFLUENCE OF MILITARY AUTOMATION	o THE FLEXIBLE SPECIALISATION PARADIGM
o BUROCRATIC AND CENTRALISED CONTROL	o COOPERATIVE AND DECENTRALISED PATTERNS OF CONTROL
o FORDISTIC PARADIGM	o SERVICE PARADIGM

It is the particularity of Europe that it is composed of very different cultures, including production cultures with strong relationships to national histories and traditions.

Therefore FAST has set up an assessment exercise to explore more in detail the factors which are fostering or hindering the implementation of anthropocentric technologies. This research exercise has started in 1990 and in one year's time results will be available.

The Tokyo workshop is part of this exercise and therefore the overall description is given below.

The FAST Assessment Exercise on Anthropocentric Production Systems (APS)

The objectives

The study aims at assessing development perspectives, during the next 15-20 years, of anthropocentric production systems. This term is utilised to describe a particular principle of technology design. Anthropocentric systems are human centred systems based on human competences and holistic organisations (low divisions of tasks, flat hierarchies): Anthropocentric Production Systems are considered to be a powerful instrument to improve the production performance and European competitiveness of Europe at the beginning of the 21st century.

The structure of the research work

The core of the research work is concentrated on two areas: the assessment of prospects of APS in all member countries, and the requirements for their technical development and diffusion.

The core activities are complemented by two further activities, a world comparison of production modes and a summer school on APS.

Fig. 1. The structure of the research work

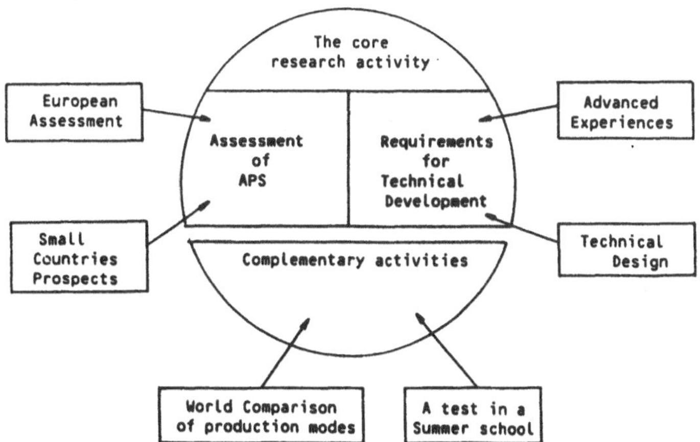

The core research activity

The *Assessment of prospects of APS* is a socio-economic exercise in eleven European Community Member Countries to analyse the different (national) production systems and cultures with the view to comparing to what extent they offer different conditions and responses to the design and development of APS. Beside such an overall assessment, a case study is carried out on three 'small' countries (Ireland, Portugal, Greece) to explore the extent to which and under which conditions anthropocentric technologies could play a role in their development process.

The *technical focus* of the core activity is split up into two parts. The first part will study selected advanced experiences with APS in different Member Countries.

The second part will be an attempt to define the technical development needs that are required for the further development of APS in Europe.

The complementary activities

Complementary to this core research two further activities are scheduled:

- *A world comparison of production modes and APS is scheduled for the USA, Japan and Europe. In this case FAST tries to valorise an existing network, CAPIRN, by expanding its activities to the consideration of Anthropocentric Production Systems*
- *A first test of the outcome of the core research will be carried out by a summer school on APS.*

Organisation and Management

• To organise the core research activity a network (see Annex) has been set up. We have chosen a main contractor for coordination and the final report (The Wissenschaftszentrum Nordrhein-Westfahlen-Institut Arbeit und Technik (IAT), Prof. Lehner). The expertise of the institute lies in both areas, in the socio-economic and in the technical field.

• The activity on *World Comparison of Production Modes* is to be carried out by the CAPIRN network.

The summer school is being organised together with the European Foundation for the Improvement of Working and Living Conditions.

• To help the Commission to implement the research activities an *Advisory Experts Board* has been set up. It is composed of experts chosen on a personal base from industry, trade unions and public administrations.

• FAST has established an in-house research capacity, thanks to three fellows who have sound experience in the research field:

- Dr. Jutta Schwarzkopf (University of Bremen, D)
- Prof.Oluf Danielsen (University Roskilde, DK)
- Roberto Benati (FIOM, I).

Budget

The budget relies on co-financement. The Commission of the EC is injecting 255.000 ECU (nearly 300.000$). The overall funds amount to 716.000 ECU (around 800.000$). The *core research activity* has a budget of 480.000 ECU to which the Commission is contributing 185.000 ECU. The largest part of cofinancing comes from the state of Nordrhein-Westfahlen with 220.000 ECU.

Final Remark

The assessment research on Anthropocentric Production Systems is under way. In parallel to the 3rd CAPIRN conference, the European networks will meet in Germany the 24-27 September 1990 for a conference *Production Technologies, Social Organisation and Competitiveness* in Gelsenkirchen where the first draft results will be available.

Annex

Detailed composition of the research teams and networks

1. Overall coordination and final report

Institut fur Arbeit und technik des Wissenschaftszentrums Nordrhein-Westfahlen (IAT)
Prof. Franz Lehner

2. The core research activity (Network I)

Assessment of prospects of Anthropocentric Production Systems (APS) in Europe and Elaboration of Requirements for their Technical Development and Diffusion

The network is composed of several research groups strongly connected with others under the general coordination of the IAT

A. Assessment of APS

• Members of the research group on the European Assessment of APS

Coordination : IAT,D

(B) P. Berckmans, E. Eysackers, Stichting Technologies Vlaanderen, Brussels
(NL) B. Dankbaar, MERIT, Maastricht
(F) D. Linhart, CNRS, Paris
(DK) L. Ramussen, Technical University of Denmark, Lyngby

(I) O. Marchisio, Studio Giamo, Milano
(UK) T. Charles, Staffordshire Polytechnic, Stoke on Trent
(E) Prof. Holmes, Barcelona

• Prospects of APS for small countries

Coordination : SUS Research, IRL

(P) Prof.A. Steiger Garcao, Universidade Nova de Lisboa
(H) Prof.Z. Papadimitriou, University of Thessaloniki

B. Requirements for Technical Development and diffusion

• Analysis of advanced experiences of APS

Coordination: HDZ, Technical University of Aachen, D

• Technical Design proposals for APS

Coordination : Dr. Paul Kidd, Cheshire Henbury Managements Consultants, UK

• Core group:

(IRL) Prof.J. Browne, University College Galway
(UK) Prof.T. Husband, Imperial College of Science and Technology, London
(D) Prof.G. Selinger, Fraunhofer IPK, Berlin

• Proposed advisory support group:

(I) Prof.R. Michelini, University of Genova
(F) Prof.G. Doumeingts, University of Bordeaux
Mr.A. Laffaille, Association francaise de robotique industrielle, Paris
(DK) Dr.I. Sejersen, Danish Technological Institut, Copenhagen
(B) Prof.V. De Keyser, University of Liege
(NL) Dr.R.G.H. van der Heiden, ITP, TUE/TNO, Eindhoven
(UK) Prof.M.G. Rodd, University of Wales, Swansea
(D) Dr.P. Brödner, Institut Arbeit und Technik, Gelsenkirchen

3. The complementary activities

World Comparison of Production Modes and APS (Network II)

To this end we got in contact with an already existing network: The CAPIRN network (Culture and Production International Research Network) which is carrying out a research project on the impact of different national production cultures on the design and development of programmable automation in the USA, Japan, Germany and Italy.

Network co-ordinator: Prof.F. Rauner, University of Bremen (D)

Principal Members of the network are:
Prof R. Gordon, University of California, Santa Cruz (USA)
Prof. Yuji Masuda, Tokyo Kezai University (JAPAN)
Prof. Roberto Camagni, University of Padua (I)

Three workshops/conferences are scheduled by CAPIRN:
- Santa Cruz December 1989
- Tokyo May 1990
- Gelsenkirchen September 1990

They will specifically address issues related to APS.

Summer school on APS, together with the European Foundation for the improvement of Working and Living Conditions, Dublin

Coordination: HDZ, D

Notes

[1] Prospects for Human Work, Industrial Organisational Strategies: The FAST II Programme 1984-1987, Results and recommendations, Vol 2, Brussels 1989

[2] *European Competitiveness in the 21st Century - Integration of Work, Culture and Technology*, by Mike Cooley on behalf of an European Expert Group, FAST, Brussels 1989.

[3] Proposal for a Community R&D Programme on Human Work in Advanced Technical Environments, FAST, Brussels 1989.

The Beginning of the Third Industrial Revolution and Changes in Industrial Sociology: Towards a Better Environment for Man

Valery K Zaitsev

It is a great honor to have been invited to speak before such a distinguished audience. Today I would like to present my views on what are the main features of the third industrial revolution, why human renaissance should be a major element of the technological development and what problems arise from cultural differences in the sphere of technology transfers.

One of the evident features of today's economic society is the shrinking time frame for change. In the past, social and economic progress was measured in terms of centuries; today, however, it is measured by decades or even weeks as in the case of the Eastern European countries. Clearly, the tempo of historical changes is accelerating. As a result the words 'The 21st century' came to imply peace, prosperity, progress and cultural advancement - the whole spectrum of popular hopes and aspirations for the future.

All countries have been actively seeking to develop science and technology as the key to economic and social development and to resolving social problems. As a result R&D capabilities in the world have grown markedly.

Value concepts relating to technology naturally differ between countries in accordance with national circumstances, national traits and market characteristics.

For example, in the Soviet Union and United States importance is still placed on military and space development technology. Strategic Defence Initiative (SDI) and the like projects in the USSR are typical in this respect. France is outstanding in nuclear power, aircraft, space and marine development. In Japan, as well known, the development of technology is centred on consumer goods, and the application of new developments to the mass production of goods is enthusiastically pursued. Thus, each nation have its special forte.

But the problem is that it used to be that military technology was synonymous with leading-edge technology, and the technology used by the ordinary citizen was considered merely a by-product. But this situation is now undergoing a drastic change, especially in the Soviet Union.

We began to realise that the developing technology for peaceful purposes is far more difficult than developing technology for military purposes. In the case of military technology the top priority requirement is to raise specific aspects of performance to the limit, and little concern is given to many other aspects.

As a result, it is possible to obtain high-level technology which is outstanding in certain points but which, seen as a whole, often lacks balance and proved to be very one-sided. In developing technology for the daily lives of ordinary people, a balance must be maintained between safety, cost, reliability, ease of maintenance and service, and handling ease.

Therefore, in the Soviet Union who started a vast program of conversion, the transfer of military technology to the civil sector does not take place readily. In 1989 the share of consumer-oriented production in the total value of production of the Soviet military-industrial complex was 40%. By 1995 it should increase up to 60%. It is neither quick nor always easy to bring technology of this ilk to the practical level of consumer technology.

America is now pushing with research into the Strategic Defence Initiative as a model national plan for the development of advanced technology. Even those allies of the United States who harbour doubts about the efficacy of SDI as a military strategy have a strong interest in the sophisticated technology that this research may produce. However, there is probably a need to examine more seriously, and from more angles, whether or not this way of proceeding with advanced technology is a wise, efficient course from the point of view of expectations of the ordinary people.

While technological innovation has been at the foundation of industrial development and economic growth in any society, it has been only in Japan and other late starters of industrialisation that government recognized the importance of national technology policies to accelerate the pace of technological innovations in the country even for nondefence purposes.

Often believed that it is a miracle that Japan has achieved such a high technological capability mainly with civilian sector efforts. In view of the country's position, Japan choice of civilian-type technological development and popularisation and socialisation of high technology proved to be a correct choice.

While the nuclear superpowers have been pouring massive human resources and huge amount of capital into the development of military technology, Japan have been directing both human resources and funds for research and development almost exclusively towards technology for general consumers. Now that this policy has come to fruition, Japan has become a nation with world-class technological capabilities in a broad range of fields.

Now all advanced countries regard the development of high technology and its industrial application as one of their top priorities from the standpoint either of economic vitalisation or economic security. This also became a matter of national prestige. The high technology trends most strongly spotlighted today are those in the fields of microelectronics-information-communications, new materials, bio-technology, new energy, and space-aeronautics.

It is understood that the industrially advanced nations are now on the brink of a third industrial revolution. In the previous industrial revolutions, technology was developed to take the place of human limbs. But attention in the new industrial revolution has turned to the development of technology to take the place of human brain.

In any age, there is something akin to a 'field of the moment' which develops very fast. Computers and communications are the favourites of this new revolution. These technologies are now making steady day-to-day advances.

As far as the main features of the new industrial revolution are concerned, we can summarize as follows:

First, the information-related technology has special properties which can be expressed with the words linking, composites, fusion and integration. Material industries like iron and steel, and energy industries like electricity, are now widely adopting information-related technology and using it to facilitate modernisation and rationalisation. Such links and composites are important themes in all fields of technology. As this happens, we will start to seek new specialists, I mean we will need 'fusion-type human resources' capable of doing a number of individual specialities. Furthermore, we cannot expect new development unless various organisations are run more openly than has been the case in the past. Consequently, in the Soviet Union, for example, little can be achieved unless government ministries and offices cooperate across the bureaucratic border lines. Perhaps, the conditions in Japan are more conductive to the exchange of different types of information than anywhere else in the world.

So one of the great advantages of the third industrial revolution is that any number of new technologies can be developed by linking together several existing technologies.

Another feature is that advanced technologies in themselves are diversifying. To express this in different terms, we are living in the age of diversification of creativity. It goes without saying that the discovery of new principles and the invention of new technology are creative activities. But finding new ways of using existing products, for example, can be equally creative. This is an age when everybody can become creative.

Human beings are diverse by nature - extremely diverse as indicated by the fact that each individual has a different genetic structure. It has now become possible to pursue and express this diversity more easily than in the past.

Now that people's desires are diversifying it is no longer possible to satisfy everybody's preference by mass-producing large quantities of the same things. Accordingly, we now need to think of turning the basic mass production set-up to manufacture a 'large-variety, small-lot' of products which can fulfil the needs and wants of all of us.

As a result, science and technology will become more complicated in the future, a special emphasis should be placed on the development of technology where cultural and human aspects will be considered as major elements. I believe that the phrase 'human renaissance' successfully expresses the goal that advanced technology should work toward.

The fruits of advanced science and technology have penetrated in every day life. They have helped overcome many of the problems faced by society including some mass diseases, food and energy shortage. This made people less conscious of science and technology itself and often consider them simply as part of everyday life.

On the other hand, disadvantages are also obvious. New technology sometimes leads to new problems including overnutrition, bioethics, computer crimes and invasion of privacy. This provokes people's fear of and disbelief in science and technology, often hindering progress in useful fields.

People are increasingly dependent on various high technology systems including electric power, gas, transport and communications. Only when the safety and stability of these systems is ensured can the people enjoy stable and comfortable lives. In the Soviet Union Chernobyl nuclear power plant explosion caused so big and long lasting damage that people have got a strong negative attitude towards nuclear energy industry in principle.

Because of this, it is important to help the people deepen their knowledge and understanding of modern science and technology and to create the information systems necessary for the people to make an independent choices.

I believe that material wealth and these new problems are providing us, particularly in the advanced countries, with a chance to consider a better environment for man in the future and the ways in which science and technology can contribute to the diversified economic, social and cultural interests of the people. Cultural values, individuality and creativity have become to be more strongly emphasized in modern society, and science and technology are increasingly expected to reflect this trend.

In recent years, the value perception of the public is shifting from material comfort to spiritual fulfilment, thereby putting pressure on researchers and planners to develop goods that satisfy the cultural and human needs of the consumer. These goals can be achieved only through the formation of high-quality intellectual stock supported by historical and culture heritage.

Contemporary science and technology are advancing remarkably and deepening their direct relationship with man in three directions.

First, the manufacture and processing of substances are increasingly being done at atomic level.

Second, intelligent elemental technology that learns from the intelligence of man, is developing.

Third, science and technology are becoming more comprehensive to grasp a human being as a total organism in its living condition. Life-support systems and bioelectronics developed directly from our knowledge of living organisms.

Fourth, progress is being made in the development of advanced technology including computer graphics and holography that can appeal to the feelings and sense of beauty.

These directions will make it possible for us to meet the advanced and subtle needs of people.

At the same time, the necessity is growing for the development of innovative technology that can contribute to the betterment of the world to win the trust of the community of nations.

The solution of the disquilibriums existing in the international society (international trade and payments imbalances between developed countries; the North-South gap, uneven distribution of cumulative external debts, etc.) should be achieved through innovative growth of the world economy.

The development of technology not only extends the new frontiers of the world economy. It is also expected to contribute, from a long-term point of view, to the solution of common problems of people, such as global environmental disruption now threatening mankind, exhaustion of non-renewable energy and other resources, and increasing stress suffered by people as a result of automation and information explosion.

In order to solve these problems smoothly and achieve constant development of human society in the 21st Century it is increasingly important to facilitate the global exchange of technology, taking into account the ever-rising cost of technological development.

But while it is said that national borders do not exist in science they do exist in technology.

It is also certain that international specialization and cooperation in high technology will not progress so readily. This is because all advanced countries are promoting and developing high technology fields as industries vital for the future invigoration of their economies, and thus there is a complicated interaction of national interests. The adjustment of national interests concerning protection of high technology and the opening of the high-tech market pose difficult problems.

The management of technology transfer is confronted by at least three problems. First is the problem of what kind of technology should be transferred. On this point, views are divided between the supplying and the recipient countries, or even within the recipient countries. Second is the problem of how to start extensive technology transfer and how to convince the supplying countries to look on this positively. In fact revenues from technology transfer are much smaller than exports of products. But one of the greatest stimuli for technology transfer would be to build a favourable international division of labour between the technology suppliers and receivers. Third, is how to help the transferred technology take root in the recipient countries and achieve the planned objective.

In essence, this is a matter of culture. Each country has a unique culture it can be proud of, and it is necessary to make efforts to incorporate this culture in products made as a result of technology transfer. Only after this is done, can transferred technology truly become the technology of that country.

Foreign economic aid and technology transfer it promotes will invariably affect the indigenous culture of recipient society. Economic development means to some degree the replacement of one culture by another. If recipient countries hope to achieve real economic gains, they must accept this. This is one of the prices they pay for progress and prosperity.

Recently, there is a growing willingness evident all over the world including the Soviet Union to try to learn from the Japanese experience. Of course we understand that many aspects of the mechanism supporting Japan's technology prowess would be difficult to apply unchanged in other countries, especially in those with different economic systems. But better understanding abroad of Japan can stimulate in other countries their own special characteristics and strengths, contributing to their own technological growth and, through this, helping to realize a more harmonious world.

Contrary to public perception, rather than being a packager of Western technology, Japan is increasingly the source of new ideas and know-how. So it is important to consider the cultural, social and historical roots that underlie this Japanese strength.

Now a new world is emerging that requires a new partnership. As the Soviet Union is concerned we need new policies adjusted to the new realities of our world today. Though the process of perestroika is proceeding with great difficulties the Soviet Union will never be the same.

There can be no denying that we now live in a global village. Our experiences, whether as individuals or as nations, determines our outlook on life. That is why we so often find it difficult to deal with change or to understand the new. But the interests of survival as well as economic security considerations behoove us to count on one another, forming bonds of mutual dependence.

Industry, Culture and Technology Transfer: Contemplation on Recent Changes in China's Consumption Culture

Feng Zhao-Kui

After the establishment of New China, preferential development had long been given to heavy industry, that is, the industrial departments producing the means of production. As for light industry, that is, the industrial departments producing the means of subsistance, its development had long remained at a level that could basically meet the needs of the people's livelihood, so that many products had practically undergone no changes in variety for decades. For example, the commonly called 'three major articles' meant the watch, bicycle and sewing machine in 1970s just as in 1950s. Till the end of the 70s, these three major articles, particularly the brand-name goods, were still rationed.

Since the policies of reform and opening to the outside world were implemented in China at the end of the 1970s, the development of light industry and tertiary industry, which may also be called the *lower reaches* industries has been accelerated, as compared with the growth of heavy industry, which may also be called the *upper reaches* industry. This implies that the focus of industrial development as a whole has, to some extent, been shifted from upper reaches to lower reaches. The development of the manufacturing of electric appliances has been particularly rapid. For example, the 'three major articles' mentioned above came to mean the TV set, refrigerator and washing machine in the 1980s, corresponding to the 'three magic devices' in Japan in 1950s-1960s. However, of these, the black-and-white TV set was soon replaced by the colour TV set, the one-door type refrigerator, by the two-door type with an independent ice-chamber, and the one cylinder type washing machine (without desiccator), by the two-cylinder type (with desiccator) and then by by the automatic one-cylinder type. In the second half of the 1980s, video tape recorder, musical components and microwave oven became more common in cities. Compared with the progressively upgrading process of the 'three magic devices' in post-war Japan (the black and white TV set, the refrigerator and washing machine in 1950s-60s, the colour TV set, the sedan and the air conditioner in 1960s-70s, and the large size colour TV set, the big sedan and the video recorder in 1980s), the consumption of some durable consumer goods among China's city inhabitants in the 1980s has witnessed a 'grade skipping'.

Reasons for the Development in China

Reasons for this rapid development of 'lower reaches' industries, such as the manufacturing industry of durable consumer goods and the tertiary industry, are as follows:

(1) Because of the opening to the outside world, the consumption culture of developed countries was rapidly introduced into China. Especially the 'high-grade' (relative to the level of consumption in china) consumer goods made in Japan and other countries have rushed into China (as seen from the incidents of speculative reselling of cars in Hainan island, some commodities entered the domestic market through unusual channels) together with the images of consumption culture spread by the advertisements. This has aroused some people to pursue the way of consumption in developed countries, which aided by the increase of their income in varying degrees since the reforms, has created a great market demand for 'high-grade' consumer goods. Such market demand naturally stimulates the development of domestic manufacturing industry in 'high-grade' consumer goods as the government has imposed certain restrictions (though with many loopholes) on the importation of 'high-grade' consumer goods for the sake of protecting the national industry.

(2) The need for developing domestic manufacturing industry of 'high-grade' consumer goods leads to the 'lower reaches' tendency of technology import. This has two implications. First,while most of the technology import from 1950s to 1970s has been concentrated on heavy industry (such as the 156 engineering items introduced from the Soviet Union and East European countries in 1950s, the petrochemical, metallurgical, electronic, synthetic fibre and precision machinery items introduced from Japan and Western European countries in 1960s, and the chemical fertilizer, synthetic fibre, steel rolling items introduced in 1970s), many imported technologies in 1980s have been for the manufacture of consumer goods produced by the 'lower reaches' industry.

Second, even within the 'lower reaches' industry, there is also the tendency towards 'lower reaches' in technology import. For example, within the manufacturing industry of durable consumer goods such as the colour TV set and the refrigerator, there is also the difference between 'upper reaches' and 'lower reaches (the manufacture of kinescope and other electronic devices can be regarded as 'upper reaches' technology and the installation of the assembly line, as 'lower reaches' technology and so on). Over recent years, the importation of such 'lower reaches technology' was much faster than that of the 'upper reaches technology' for manufacturing devices and materials. According to statistics taken at the end of 1985, 113 colour TV assembly lines, 70 refrigerator assembly lines, 15 copy machine assembly lines, 30 foreign clothes production lines, 7 laser phonograph assembly lines and 15 camera production lines have been imported to china. The excessive import of assembly lines has led to a shortage in the supply of home-made devices and materials and dependence on imported goods, and in the case of insufficient foreign currency for imports, many of these assembly lines have to suspend production.

Apart from the assembly lines of durable consumer goods mentioned above, many cities of provinces have vied with each other in importing many beer pouring lines, bread production lines, fast food production lines, macaroni production lines, bean curd production lines etc.

(3) With the support of our government, the tourist trade has made rapid progress and become an important factor propelling the development of the 'lower reaches industry' as a whole.

The fast expanding tourist trade is now a major source of foreign currency and has promoted the development of the catering trade, commerce, building industry, communications and transportation, the manufacturing of arts and crafts and special or indigenous products etc. The growth of the tourist trade has facilitated international contacts and cultural, scientific and technological exchanges, which are 'super-economic results' incalculable in terms of money.

However, there was also a certain degree of 'over heated' growth in tourism in recent years. In some major cities, the construction of tourist facilities tends to be excessively large in size and the standard is too high. In Beijing, about 40 thousand new guest rooms will have been completed by 1990, which in addition to about 17 thousand old guest rooms, will make a total of 57 thousand, 30% more than the estimated need at that time. In Guangzhou, the total number of guest rooms reached 59 thousand in 1986, exceeding that of Hong Kong, but the construction boom is still high. In fact, surplus tourist facilities, designed for foreign guests, have already been used to receive domestic guests, resulting in the quick spread of the luxury hotel culture of the developed countries in Chinese society.

Some say, in China, our tourist trade achieve economic results through the 'exportation' of natural scenery. But this is only one side of the question. The other side of the question is that our tourist trade also plays the role of 'importing' the 'social scenery' of other countries into our country, providing a 'demonstration centre' for high consumption in developed countries. Particularly because some people persistently seek high-grade and luxury tourist facilities and services, the high consumption culture of capitalist countries, far above the consumption level of the ordinary Chinese, has been introduced into China.

(4) China built the Third Front in its southwestern part in the 1960s and Special Economic Zones along it's Southeastern coast in the 1980s. The former is of the closed type, the latter, the open type. During my recent investigation tour in Shenzen and Zhuhai, I found that the Special Economic Zone is, in certain senses, a 'large shop window' for the high consumption culture of developed countries, showing a quite comprehensive image of it, so people from the interior often regard their trip as a 'quasi-visit abroad'.

Of course, the Special Economic Zone is primarily a base for introducing the production accomplishment of developed countries. However, the proportion of 'lower reaches' or consumption industry is comparatively high in the imported industries. In addition, as the Special Economic Zone can more freely take in the consumption culture of developed countries, it pushes forward our 'lower reaches' industry both in industry and in culture.

As the consumption culture of developed countries, symbolized by the Sedan, electric appliance and tourist service, rushes into my country, it must be subjected to a sieving according to the conditions in China and adopt a pattern peculiar to China in the primary stage of socialism. It has the following characteristics:

(1) Durable consumer goods, such as the colour TV set, refrigerator and washing machine, which are not so expensive (about one or two years' income of an ordinary salary man), are rapidly becoming popular. One reason is that the three major articles meet the pressing need of the masses of rural and urban people to improve their material and cultural life. Another reason is that the price of these electric appliances is within or not so much beyond the actual ability of the ordinary people to pay. While the housing reform is not yet in full swing and the city inhabitants are still relying on the government to solve their housing problem, to purchase the 'three major articles' becomes the main 'capital construction' for ordinary families.

(2) Durable consumer goods of the information type, such as the TV, radio (especially short-wave radio), tape recorder and video tape recorder, are becoming increasingly popular and the 'grade skipping' phenomenon occurs, for the video tape recorder, for example, popular in developed countries only in recent years. This fact reflects the people's

yearning for knowledge and information. The popularization of TV, which can spread knowledge and information graphically, thus breaking through the 'bottle neck' of low literacy rate, is very important for the modernization of China where the illiteracy rate stands above 20% of the population.

(3) Popularization of private cars is extremely slow. Now the proportion of sedan owner to population is 1:1.9 in United States, 1:2.5 in Western Germany, 1:4.5 in Japan, 1:26 in Soviet Union, 1:16 in Malaysia, 1:94 in Pakistan, 1:10200 in China. It is estimated that private cars will provide special convenience to only a small minority of people for a long time to come.

Compared with the electric appliances industry, growth of the automobile industry depends more on developing a series of basic industries, and is more restricted by the comprehensive development of the country's industrialization. As a consumer item, the price of a car is far beyond the whole life's income of an ordinary salary man, so it will be impossible for ordinary people to own private cars in the foreseeable future. Besides, the development of highway communication could not bear the consequences caused by a sharp increase in the number of cars should it happen.

(4) The over-development of luxurious hotels and restaurants. It has taken place mainly in the name of attracting foreign tourists, but it has also resulted in satisfying the pleasure-seeking desire of small number of people including some officials who can use public funds to enjoy themselves. The emergence of such tourist facilities embodies a combination of the gourmandism culture of China's feudal ruling class and the capitalist hedonist culture. In some cases, as people say, life is more extravagant in poor countries than in rich countries.

(5) The accelerated development of private telephones. Until now, the popularization of the telephone in China is slower than not only the developed countries, but also the average developing countries. However, due to the progress of information technology and the urgent need for information, and because of the relatively small investment and quick recovery rate, the number of private telephones increased very rapidly in recent years (about one million per year). It is estimated that the radius of action of the Chinese will be extending faster in information space than in physical space.

To state succinctly, the development of the consumption culture in Chinese society will be centred on electric appliances, and not on private cars; will lay more emphasis on extending the radius of action in information space than in physical space. Such a development pattern, is attributable to many factors, including the conditions in China basically characterized by it's biggest population in the world, the development level of China's industrialization, and the effect of informationalization in Western countries.

Positive Effects of the Chinese Consumption Culture

The spread of consumption culture and consumption industry of developed countries into China has complicated repercussions in Chinese society, which has been closed to the outside world for many years. The positive effects are:

(1) It makes the short term target of China's modernisation development, i.e. to lead a fairly comfortable life economically, a rather concrete and graphic concept.

Now it is certainly no easy job to purchase the 'three major articles' and other electric appliances with the average income of the Chinese people. This fact leads to the formation of a reasonable, appropriate target pursued by people in material life, which helps enhance their enthusiasm for work. One of the objectives of economic restructuring in China is to break the *large canteen cauldron* and the *iron rice bowl* and hence to form a mechanism, under which 'he who makes greater efforts, gets higher remuneration'. This mechanism needs the coordination of attractive images of the consumption culture, so that the 'higher remuneration' gets concrete material embodiment, otherwise, people may think that there's no need to earn more money. In this sense, to introduce the healthy, reasonable parts of the consumption culture of developed countries has a positive effect on our economic reforms.

(2) Introducing healthy, reasonable consumption has changed, to a certain degree, the concept of 'paying attention to production and thinking little of consumption', and has made people aware that healthy, reasonable consumption is also a kind of 'production', that is, the 'reproduction' and 'expanded reproduction' of people's labour vigor and working ability. For example, watching TV enriches the labourers' knowledge and widens their vista; the time saved by using the washing machine can be used to learn culture and enjoy their spiritual life; the refrigerator is good for reasonably arranging daily meals and improving the health of labourers. On the other hand, the great market demand created by the rising level of people's consumption and the huge population of China will surely give a strong impetus to the economic development.

(3) As mentioned above, the popularization of 'information type' durable consumer goods, such as the TV set, tape recorder and video-tape recorder, is of considerable significance to developing education of the whole people, and to raising their average cultural level.

(4) Where there are commodities, there is a culture behind them. Following the importation of durable consumer goods, which could be regarded as the 'hardware' of consumption culture, the 'software', that is the culture and ethics of the capitalist countries has also spread into China. People, who eat beef and mutton, do not necessarily become cattle and sheep. The healthy and useful parts of capitalist culture and ethics could be used as important nutrition for developing socialist culture and ethics.

(5) Although we have taken a tortuous path in technology import, our practice in recent years has further testified to the necessity of importing advanced technological culture. The world is advancing in an unceasing flow of civilisation. The 'highland' of civilisation formed in one historical period is sometimes shifted to other places owing to the 'crystal movement' of civilization. Therefore, it is undesirable to cherish a 'sense of centre', ever posing as 'an old and great nation of civilization'. In this sense, actual contact with the consumption culture of developed countries has contributed to increasing one's modesty in eagerly absorbing the advanced culture and science and technology of other countries.

Negative Effects of the Chinese Consumption Culture

Of course, from the actual conditions of introducing the consumption culture from developed countries, you can see some negative effects:

(1)The 'high-grade' consumer goods were first produced by the developed countries, especially in the United States. As for Japan, which has attached importance to civil technology, as a proverb says 'the late-comers surpass the old-timers', it has become the biggest producer of electrical appliances and cars in the world. It is obvious that in the USA and Japan, the technology and production of consumer goods comes earlier than the consumption, hence the consumption level rises gradually together with the production level. However, for the people of developing countries just opened to the outside world, they sometimes see first the consumption in developed countries and then the production behind it. It could be seen that in recent years there was an excessive eagerness in importing and publicizing consumer goods and the way of consumption in developed countries. Not only cars and electric appliances, but also cigarettes, wine, beverage, coffee, perfume, toiletry, clothes etc, have been mass-imported from foreign countries. For example, in recent years my country paid above 199 million dollars annually for imported cigarettes. In Shanghai a young father said:

> *My son never uses domestic goods except water, air and fruit. All that he eats and uses are imported goods.*

In contrast, there was less enthusiasm in importing and publicizing the production culture. The rise of production efficiency is relatively slow. It seems that it is much easier to learn consumption than to learn production. There is the so-called 'unduly high consumption' phenomenon in China, meaning that the level of consumption is being raised with a rhythm quicker than that for development of production, although the actual consumption level of the ordinary Chinese is still very low.

(2) When the consumption culture of developed countries, whose per capita GNP amounts to ten or twenty thousand US dollars, has spread into developing countries, whose per capita GNP amounts to several hundred dollars, it is difficult to avoid the widening gap of consumption between people with different levels of income. In addition to the problem of unfair distribution, the widened gap of consumption could lead to spreading the 'sense of unfairness' in the whole of society and even affect stability. During my recent visit to Shenzen, I was surprised to find that an entrance ticket to Kara OK costs 150 yuan (Renminbi), almost my monthly salary as a professor. Although those working in Shenzen get higher salary than their counterparts working inland, I think, after all, only a few would pay 150 yuan for a Kara OK ticket. I got acquainted with a self-employed worker on the plane, who told me that he had a private car, and the expenses for using the car (including gasoline fee, road toll etc.) totaling several thousand a month, the sum of my salary for years. The items of consumption he chose are the luxurious sedan, precious jewellry and travelling abroad, all much higher level than the 'three major articles' for the ordinary Chinese. I don't want to deny that an income gap is favourable for increasing economic vitality, but too large an income gap may adversely affect economic development and social stability.

(3) In the capitalist consumption culture, there is a tendency to be rich materially and to be poor spiritually. Therefore, if we cannot develop socialist culture and ethics simultaneously with the quick introduction of capitalist consumption culture, it is hard to avoid the imbalance between material consumption and spiritual consumption. For example, although there is a certain popularization of video-tape recorders in cities, in many families these appliances become decorations for lack of adequate supply of 'software'. Recently, the publishing business was slack in China. Many valuable cultural and academic works have

been cut down in the editorial process by the publishing houses now in a predicament. This is also an indication of the imbalance between material and spiritual consumption.

(4) About two years ago, when nude art exhibitions were held in Beijing and other major cities, people of all age levels, from teenagers to elders of 60s or 70s, would form a long queue in front of the exhibition halls to gain entrance. This points to the great shock brought by the introduction of Western culture and art into Chinese society which has long been closed to the outside world. Although there are still different evaluations of the nude exhibitions and the like, it can be said with certainty that at one period some unhealthy, morally contaminating capitalist culture has spread into China. For example, the imported video-tapes and publications with pornographic contents or advocating violence have played a role of corrupting social moral and are harmful to the mind and body of teenagers. In capitalist countries, great importance is attached to freedom of the press, but where there is freedom for a few to publish pornographic literature, there is no freedom for the vast number of parents to protect the healthy growth of their children. The anti-pornography campaign launched in China last year precisely aims at restricting the freedom of a small minority, and protecting freedom of the great majority.

(5) Although it is a good thing to popularize the use of electric appliances, in out country it is necessary to impose certain limitations on the use of high power consumption type electric appliances such as the electric cooking utensil and large size refrigerator because of the lack of electricity to support fully the development of industry and economy. From the long-term point of view, if we popularize such appliances, or even private cars without restriction in our country with its population of 1.1 billion to the level of developed countries in per capita power consumption, the power consumption of China would increase several dozens or even a hundred times. Such an increase, realizes on the basis of an energy structure with coal as its main part, would seriously affect the global environment.

Defects in Traditional Capitalist Industrial Civilization

In my opinion, problems arising from the introduction of the consumption culture of developed countries are a reflection of the defects inherent in traditional capitalist industrial civilization. These defects are as follows:

(1) The widened gap of consumption between the Chinese people of different levels of income in recent years is a symptom of the domestic trend of North-South polarization which has become a very serious contradiction in traditional industrial civilization. This is because the traditional industrial civilization has created a polarizing, unstable world with per capita GNP differing considerably, in some poor countries a special social stratum has emerged, which is eager to emulate the material benefits enjoyed in rich countries, in spite of the fact that the per capita GNP of rich countries and poor countries is respectively at the upper end and lower ends on the international economy scale. China, as a socialist country, obviously must avoid the rich-poor polarization while opposing equalitarianism and the 'large canteen cauldron' in the process of economic restructuring.

(2) In a certain sense, the 'consumption craze' which occured in recent years in China could be regarded as a social resilience after the long stagnation in consumption level. On

the part of consumers, however, the 'consumption craze' is influenced by the belief 'the more consumption, the more honourable' prevailing in industrialized countries. This belief prompts people to pursue a high consumption far exceeding the essential needs of life, thus giving rise to a society of extravagant consumption.

(3) The imbalance between China's material development and cultural development in recent years is, of course, related to the backward state of educations and low cultural level of the people, but, at the same time, this phenomenon is also related to the tendency of traditional industrial civilization to put undue emphasis on material development and ignore spiritual progress.

(4) The extravagant consumption culture of traditionally industrialized countries also influence Chinese society. In those countries, the tendency to regard excessive consumption and extravagance as an honourable thing disregards the limitation on global resources and environment.

Concluding remarks

The globe has already been 'overloaded'. Obviously, if all the people of the world were to pursue extravagant consumption as do the developed countries, global resources would soon be exhausted and the global environment broken. Therefore such a consumption culture in a certain sense can only be regarded as the consumption culture for the privileged, which cannot be spread to the whole world.

Today, when a majority of developing countries have embarked on the road to industrialization, it is very important to draw a lesson from the traditional industrialization and to lead the developing countries onto a new, healthier and more reasonable road to industrialization. This is a practical task concerning the future of not only the developing countries but also the whole mankind.

For the healthy development of human civilization, it is hoped that the developed countries will pay more attention to the following tasks:

(1) To build up a simple and frugal society, develop a rational, healthy and moderate consumption culture, and set up an industrial, cultural image which is more in conformity with the fundamental interests of the long-term development of the world as a whole.

(2) To actively transfer to developing countries various new technologies which may replace the traditional technology of high power consumption and high contamination as well as advanced technologies that are energy-saving or highly efficient in using energy and beneficial to environmental protection, taking such technology transfer as part of their own interests.

(3) To bring into play their leading role in the construction of global infrastructure in the fields of information, communication and transportation, and environmental protection. For example, to speed up the establishment of a global information network by optical fibre and satellite.

(4) To use part of the funds saved by disarmament, especially between developed countries, for global environmental protection.

As for the developing nations, it is necessary to avoid blindly copying the Western civilization and earnestly formulate a modernization strategy in conformity with conditions in their own countries.

Section II:

New Technology and the Shaping of World Economy and Industrial Cultures

The New Shape of Industrial Culture and Technological Development

Mike Cooley

The year 2000 marks the end of an extraordinary millenium. During it, humanity has witnessed the decline of feudalism, the growth of capitalism, the collapse of state Stalinism, the emergence of Cartesian science, the concentration of populations into modern cities, the development of 'earth shrinking' transport systems and above all, and for the consideration of this symposium, there has been the emergence of industrial society.

Rate of change

The last century of that millenium has been characterised by a convulsed and exponential rate of technological change, in which our speed of communication has increased by 10^7, of travel by 10^3, of data handling by 10^6. Over the same period, our depletion of energy resources has increased by 10^4 and weapon power by 10^7.

We should reflect upon the beauty and the devastation we have wrought on our two-edged way to the 21st Century. The delinquent genius of our species has produced the beauty of Venice and also the hideousness of Chernobyl; the linguistic delights of Shakespeare and the ruthlessness of British imperialism; the musical treasures of Mozart and the physical horrors of Bergen Belsen. The caring, medical potential of Röntgen's X-rays, and the horrific devastation of Hiroshima.

We have seen the polarisation of wealth and activity. In developed countries there are computer programmes to help people diet, whilst out of the 122 million babies born in developing countries in 1982, 11 million died before their first birthday, and a further 5 million died before their fifth birthday.

Masters of Nature?

At the end of the millenium, we appear to stand as the masters of nature. We scrabble five million tons of material around each year, and in doing so we shift the equivalent of three times the sediment moved each year by the world's rivers. We mine and burn billions of tons of coal each year, so venting the waste, including carbon dioxide, the principal contributor to the greenhouse effect. Our agriculture, now carried out on an industrial scale, is performed in such a way that we cause the erosion of 25 million tons of soil each year which is 0.7% of the total arable soil which has taken thousands of years to form. We put down 30kg of fertilizer per person each year to increase the crop yield, so polluting the very water that we drink.

If we continue in the present manner, we will reduce by 50% all the species of flora and fauna in less than two centuries. In fact, this is likely to be a matter of decades, taking into account the greenhouse effect. This will constitute a terrible reduction in bio- diversity, and is also being accompanied by a reduction in diversity amongst ourselves.

Among the global issues confronting industrial society today, two are particularly pressing. They are: 1. *Resource depletion* and 2. *Environmental changes brought about by human activity.*[1]

Producer and Consumer

In consequence of these issues, it will be necessary to change industrial cultures in order to ensure that those involved are concerned not just about first order effects of their work, but second and third order multiplier effects. Furthermore, since the impact of many of the new technologies tends to be all pervasive, human beings increasingly will transcend the old industrial culture which separated the individual as producer from the individual as consumer, and there will emerge a more holistic industrial culture which will link the issues with how people work and what they produce with where they live and the quality of their lives both in an environmental and psychological sense.

Industrial and productive cultures have moved from long-term cycle thinking to short-term. For example, the building of a mediaeval church would have taken (in some cases) two or three centuries. Many of the great Victorian industrial installations are still functioning perfectly, whereas modern fixed capital is obsolete and written off in about two or three years after it is introduced. New industrial cultures will require thinking in longer term developmental cycles, rather than in short term, profit maximisation ones.

Sense of Responsibility

Some professional cultures have insisted upon a professional sense of responsibility. This, in the case of medicine, goes back as far as the Hippocratic Oath. Industrial culture on the other hand, certainly since its dominance by the military/industrial complex, has resulted in hierarchal command structures in which those who work in the industries are required simply to do 'what they are told'. This is unacceptable, and will increasingly be so. External pressures will require these changes in concepts of responsibility, but so also will the influx of better educated young people who are increasingly questioning the whole basis of industrial society.

Thus the advent of the 21st century will provide a powerful psychological stimulus for re-examining the whole basis of our industrial culture and development, and hopefully, with imagination and ingenuity, we can build upon that which is best from our past, whilst identifying and gradually removing those negative and dangerous tendencies.

Evidence abounds to demonstrate beyond reasonable doubt, that there is a growing informationisation and intelligentisation as Masuda has described[2], taking place in advanced industrial societies. Before dealing with the cultural and other aspects which arise from this, it is important not to forget the industrial activities which run in parallel with these new technologies. It is frequently forgotten that you cannot fly the Pacific on a chip; you cannot make steel or bake bread in a chip, neither can you drive around in one. Information-type technologies may control the means by which these things are done, but the primary technologies themselves require careful consideration in light of the growing environmental and ecological multiplier effects.

New Materials and Products

It is necessary to consider the materials we use and the products that we make in a new light. For a sustainable future, which nonetheless is 'high tech', it will be necessary to devise processes which involve the lowest possible release of eco-impacting material to the environment, and the highest possible degree of recyclability. In future, it will be necessary to view the development of new materials in the wider context of environment, energy, health, safety and the forms of skill and means of production available in given communities. These should be seen in the context of developments which are sustainable in the long term. In the case of the material sciences, this will mean embodying these concerns at the level of Solid State Physics Research. The funding policies and research programmes should reflect these new objectives.

In parallel with this, there needs to be a development of new product ranges which accord with this new cultural outlook. Product Innovation Programmes will require changing organisational forms so that large numbers of people can be involved in using their creativity to devise new products and services.

Technology exchange mechanisms are already being developed in Europe. The SPRINT programme provides for such exchanges across the various nation states. It concentrates on providing products for small and medium sized companies. Some of the programmes involved, such as the technology exchange, tend to concentrate on products which are socially useful and environmentally desirable. Information on these new products, and access to expert advice on the best means of producing them, has been advocated in the recent FAST Report (See note 1). Product Databanks have already been structured, and can be accessed by means of production and skill, rather than in the more outmoded sector strategy approach.

Need for Diversity

In order to stimulate and encourage this development of new products, it is vitally important that regional educational and historical diversity is respected and enhanced. Differences in industrial cultures will be important in these processes. In the United States for example, there has been a tendency to emphasise economy of scale, with concentrations on large batch production of products within a Tayloristic framework. On the other hand, the European culture has generally tended to emphasise collaboration and variety. Of course, all cultures have their strength and weaknesses. Europe's lack of focus in the past has been a considerable liability. However, in light of the new requirements (some would refer to them as post-industrial requirements), those European attributes which in the past have appeared as a liability, may well begin to be a strength. It is fairly clear that cultural sophistication and creativity in the wider sense will be advantages in the coming period. It is for these reasons that the FAST report emphasises 'that Europe must not become a melting pot as the United States has been. Its regional and cultural variety will provide the basis for addressing diversified markets, and responding to the demand for product variety. This could not only enhance Europe's international trade and improve its competitive position. Means of production and product ranges, which improve the quality of life, safeguard the environment and minimises the waste of materials and energy, will not only be vital for European society but also have commercial significance for foreign markets as these issues become a growing world wide concern'. In this context, regional culture and variety should be perceived as a strength not a weakness. This runs counter to one of the main tenets of the Western technocratic outlook, which is the notion of the 'one best way', a fundamental idea also inherent in Taylorism.

At the macro level for example, the notion of the world car is seriously flawed. We can hardly imagine that a vehicle suitable for the Nevada deserts is also ideal for the snow-laden forests of Northern Sweden.

Economy of Scope

For good economic and environmental reasons, any products in future will have to be seen as appropriate for regional requirements, cultural aspirations and environmental appropriateness, so that products suit the actual requirements of the users, rather than fit in with some preconceived notion of the means of mass production. Because of this, we can anticipate shifts from economy of scale to economy of scope, with batch size going down and variety size going up. It should be fairly self-evident by now, that society cannot go on producing ever increasing numbers of throwaway cars. Already, many European cities ban cars from their centres. The chairman of Volvo was recently reported as saying that cars will soon be prohibited altogether from many cities. As a result of this, many innovation programmes are now advocating quite different forms of cars such as low energy urban cars[3].

These considerations will also have to give rise to an engineering design culture in which we think of products for their use value, not only for their exchange value. This will mean questioning many of the given assumptions of the automotive industry itself, and the industrial culture to which it gave rise. More than any industry, it fostered Taylorism whose founder said:

> *In my system the workman is told precisely what he is to do and how he is to do it, and any improvement he makes upon the instructions given to him is fatal to success.*[4]

Mechanistic Culture

This grotesque philosophy applied not just to the nature of the product and the means of production, but even to the production of people itself. Kraus formulated the problem of this century in the following quote:-

> *Henry Ford recently donated 100 million dollars for the erection of a school which he called the school of the future.*
> *'I have manufactured cars long enough', he declared, 'to the point where I have got the desire to manufacture people. The catchword of the day is standardisation"*[5].

That mechanistic culture, today permeates every aspect of our thinking to the point where there are now models of universities as factories, in which the students are referred to as commodities, the examinations as quality control procedures and the graduation as delivery. The last thing we shall require, as these environmental and other problems bear in on top of us, is systems which produce these clone-like, standardised thinking people. Such Tayloristic thinking is now recognised, even in the United States, as being counter productive, and a recent MIT report pointed out that it is a mistake to view people as a liability. Rather, they should be perceived as an asset.

Educate Rather than Train

Accordingly, a cultural outlook which is concerned with education in a holistic fashion, is likely to be much more effective than narrow system-specific or machine-specific training. In fact my hierarchy of verbs would be: you programme a robot, you train a dog (or possibly a soldier), but you provide educational developmental environments for human beings.

In Britain, where they have concentrated on training in an increasingly Americanised fashion, the consequences of this are now evident in the downtime of advanced systems. Bill Ford has pointed out that whereas in the west one concentrates on 'on-the-job training', in Japan they concentrate on 'on-the-job-learning'. He suggests that the great international strength of the Japanese is their 'balance of skills', and says 'Japan has prospered by the development and participation of a highly educated, skilled, adaptive, flexible and innovative workforce'. He suggest that the Japanese concept of skill formation embraces the ideas of education, training, experience and personal development. 'It is a holistic concept. IT does not fragment human development in terms of the vested interests of traditional institutions'. Such a cultural outlook is also more evident in West Germany and in many parts of Europe.

In consequence of that, it was held that instead of accepting that there is just one kind of production technology, the Fordist one, we should consider forms of technology which accord with geographical, cultural and environmental needs. As a result, the EEC supported a project ESPRIT 1217, in which the proposers pointed out 'technology should be regarded as part of culture, and just as cultures produce different languages, different music and different literature, why should there not be different forms of technology also'?

The result is a computer integrated manufacturing(CIM) system, now being tested out at large European companies such as Rolls Royce and BICC. In this system the human being handles the qualitative subjective judgements and the system the quantitative elements. The basis is laid for testing in practice, concepts of human centred systems or, as they are now more widely described, anthropocentric systems. Such technologies could lay the basis for customer-bound, short batch, flexible forms of production, which will accord with the growing needs of a society which must increasingly conserve its energy and materials. Such systems will also value and build upon, the most precious asset society has, which is the skill, ingenuity and creativity of its people. This form of systems development can be applied more widely than in manufacturing, and examples have already been cited in the field of Expert Systems for medical applications.

Tool or Machine?

The object in all these instances has been to provide powerful tools which support human competence, rather than machines which objectivise them. This is the important distinction that Heidegger made[6].

Underlying all of this is the need for an industrial culture which will cherish and build upon people's imagination and creativity. Historically, people believe that creativity applies to artists, writers and dramatists, but frequently forget that imagination has always been a hallmark of the great innovators in engineering and in science. The great Einstein said on one occasion 'imagination is far more important than knowledge'. Imagination is so often stimulated by the wider culture from which we spring; the festivals in which we participate; the folk tales which we are told; the poetry which we read and the dramas for which we are an audience. This highlights the need for a merging of both arts and science subjects, to nurture human beings who are sensitive to human needs and suffering, who are conscious of the impact of our work on the environment and the beauty and magnificance of nature itself. This will require holistic forms of education, and cultures in which people are expected to be concerned about what is happening to the environment and to fellow human beings.

Anticipatory Democracy

It will mean encouraging engineers and scientists to engage in what I call 'anticipatory democracy', where democracy is not just seen as a reaction to something that has already happened, or as the selection of a 'wise person' every four years to make all the decisions, but the participation of proactive, engaged members of society.

Our science and technology is now awesomely powerful. If misused, its power can probably mean the end of humanity itself, and has already done terrible damage to the environment. But that power of science and technology and productive ability, properly used, could lead us into a 21st century which would lay the basis for a cultural and industrial renaissance throughout the world, a form of renaissance which respects and builds upon the diverse cultures that are the basis of the richness throughout the world.

Notes

[1] *European Competitiveness in the 21st century: The Integration of Work Culture and Technology*, a 95 page report for EEC/FAST. Available free from FAST, Rue de la Loi 200, B-1049 Brussels Belgium.

[2] Masuda, Y (1990) Technological Information and Information Technology in the Information Society. In: Göranzon & Florin (eds) *Artificial Intelligence, Culture and language: On Education and Work*. Springer, London.

[3] Fletcher, R (1990) The greening of the motor car. In *INNOVATION* No.5 1990 LICT. PO.Box 919, Lofting Rd, London N1 1XL.

[4] Cooley, M (1987) *Architect or Bee?: The Human Price of Technology*. Chatto & Windus, London. Japanese edition: Translation Prof. F. Satofuka. Sagami Women's University, Tokyo 1989.

[5] Kraus, K, cited in Stieg: "I have no idea where I am going, so to make up for that I go faster". In Göranzon & Florin (1990) (See note 2).

[6] Ehn, P (1989) *Work oriented Design of Computer Artefacts*, Arbetslivscentrum, Stockholm

The Coming New Age of Politico-Economic Interdependence in the Global Economy

Alexander Dynkin

The close of the 20th century will enter the history of civilization marking the beginning of a new stage in technological revolution characterized by orientation to qualitative parameters of economic growth, resource-saving and protection of the environment, satisfaction of material and spiritual needs of every man.

Developments of technologies, material, technological progress most vividly corroborate the concept of progress as a forward movement from simpler forms to more complicated ones, from primitive instruments of labour and mechanisms to the most perfect types of machines. The essence of technological progress which consists in the materialization of the achievements and knowledge of science in the elements of productive forces is becoming more apparent in the course of the progress of mankind.

The emergence in the European region in the 17-18th centuries, when the first industrial revolution was prepared and developed, of civilization of a particular type changed the centuries-old character of development. Techno-geneous societies are characterized by an essentially different type of social dynamics and display great ability to progress as the two centuries' experience has shown.

Developments of the past decades show that modern civilization, having created mechanisms of accelerated advance of science and technologies, also provided a fast rate of social progress in many countries. But just when those processes attained a zenith assuming the features of technological revolution it became absolutely obvious that a certain metamorphosis had taken place: material production, science and technology which ideally are to serve mankind began first to be considered as the end in itself and then to threaten the very existence of mankind. It would be of no avail to enumerate all the negative consequences of technological progress - they are very well-known. Many scientists and public figures are of the opinion that if the existing basic trends in technological progress continue to develop, disastrous consequenceswill follow.

It is obvious that negative and positive consequences of technological progress are interwoven and it would be impossible to 'call off' development of certain technologies without destroying civilization as a whole. The efforts of mankind are aimed at finding a reasonable solution to the problem of facilitating the progress of civilization along the road of technological progress preserving the human in man and the natural in nature.

Humanization of technological progress and endeavours to solve universal problems are to become the basic characteristic, the content of technological progress. Under the impact of the globalization of economic, political and military development, industrial culture of mankind undergoes a transformation. Therefore, it becomes expedient to study, within this context, civilization with a new face, with a free individual in the centre. We are already witnessing the formation, on the basis of such principles, of the system of control over the technosphere and the ecosphere. Priorities are being established in new technologies and in R&D: Technology and the law; science, engineering and democratic transformations in the society; technological culture and systematic education; informatization and social progress; artificial intelligence and socio-economic potentialities and consequences of its utilization; the history of science and engineering as a civilization phenomenon - all this is only a rough list of problems reflecting the present and prospective realities of social life.

As regards the current stage of technological progress internationalization can be considered as the principal political and economic characteristic of its world economic aspect. Firstly, internationalization manifests itself in the identity of the economic content, trends and consequences of technological development in three centres of modern capitalism: the USA, Japan and Western Europe. Secondly, in the '80s the levels of economic and technological development of the leading capitalist countries tended to become relatively even. Thirdly, no matter how great the national technological potentials are, none of the countries would be able, on its own, to completely satisfy its demand for R&D information, high technology and machinery and to manufacture all the needed high-tech products which are becoming more and more sophisticated.

All these circumstances make for intensification of the processes of the international division of labour, specialization and cooperation in research and development and introduction of innovations, enhancement of exogenous factors of technological development and economic growth of states. At the same time internationalization entails exacerbation of international technological rivalry both due to expansion of its sphere in international economic relations and to an increasing number of its participants.

Under the impact of the above mentioned factors in the aggregate, the developed countries redoubled their efforts to elaborate a complex strategy of technological development. Despite internationalization of technological progress and evening up of the levels of technological development these strategies retain their national specific features. Striving for complete realization of national peculiarities and the countries' distinctions, as well as for making up for all shortcomings ensuring from these distinctions is an essential component of those efforts.

The economic, ecological and military context of the current stage of technological revolution expands the sphere of mutual interests of individual countries, and objectively facilitates the intensification of interdependence and integrity of the present-day world. Long-term technological trends and economic conditions, and interstate unification of social needs which are satisfied as scientific and technological achievements are disseminated imply the growth of preconditions for joint actions. Notwithstanding the unevenness of these processes a tendency toward internationalization is the leading one and it leans upon the universality of the activities of science and technology, a growing role of the basic scientific knowledge, the common inner logic of scientific cognition, and changing science paradigms.

Under these conditions intellectual products are becoming the most attractive items of world exchange. The share of high-tech products in world economic relations and international economic structures is growing. Tendencies toward the international division of labour on the one hand and integration on the other hand are intensifying. One can say that a new world entity of transnational economic potential is being formed.

But, regrettably, the gap in realising such potential is growing: some countries are far ahead, others still remain on the periphery of technological progress. This disbalance constitutes a danger for the world. Now the question is how to create conditions that would offer equal opportunities for all countries rather than to ask how the strong are to help the weak.

International cooperation, coordination and mutual enrichment of the technological conditions of different countries are important factors in the advancement of all nations and states. All countries need new technologies, notwithstanding the levels of their economic or scientific development. At the same time quite a number of problems, such as the creation of alternative, economically efficient, energy sources or the improvement of the environment, cannot be solved by individual countries on their own, however developed they may be.

The technosphere created by the scientific thought poses a threat to the biosphere. However, it is this thought which can save it. And since the biosphere has no boundaries it is necessary to go into action beyond the frontiers, to put together the available potentials, opportunities, resources and ideas. The common strategy of working against ecological disaster is needed.

With regard to these present-day realities the Soviet Union has taken the course of integration of its economy and science into the world economic system. We possess a tremendous scientific and technological potential, our achievements in peaceful development of outer space and in basic research are noteworthy. However, although able to make scientific discoveries we lag behind the West in the ability to utilize technological achievements for the good of society. This is one of harmful consequences of a protracted dominance of administrative methods which fettered initiative and entrepreneurship, of excessively rigid centralization in planning and management.

One of the aims of perestroika is to liberate the energy of science to give it free play and to pave the way for mutually beneficial exchanges.

But not everything depends on us. The hummocks of the cold war still stand in the way of large-scale technological cooperation. There are a number of Western countries that still perceive scientific exchanges mainly in the light of 'position of strength' stereotypes to the effect that the gap in defence technologies should be maintained. Such countries nurture plans to win the arms race, to exhaust their potential partners economically.

The 'Berlin Wall' which divided the world on the basis of ideological differences has been destroyed, but in the sphere of science and technology there such a wall of division and discrimination still remains - the COCOM.

The age of global challenges demands global conversion of thinking and policies. In world affairs we are all learning how to regard the interests and concerns of each other and we should learn to do the same in the sphere of scientific and technological cooperation; and if our partners are apprehensive that technological exchanges may serve military purposes ways must be found to allay these apprehensions. It seems expedient to work out reciprocal approaches, to find out mutual concerns rather than to stick to different lists of restricted goods. It is necessary to instill confidence in our partners that technological exchanges will answer their purpose. For our part we do not want new technologies to undermine our security either.

Establishment of the systems and structures of technological confidence should be based on openness and verifiability of cooperation in this area, as well as above conditions on the utilization of its results. We can see the first sprouts of such thinking in the nuclear sphere, for example. I would remind you that the guarantee of nouse of atomic energy for military purposes was taken as the principle of the IAEA Charter in the '50s. It is possible to frame a concept of open laboratories and to put it into practice, to expand the scope of joint R&D, to think out how to organize field inspection of the utilization of the transferred technologies.

In conclusion I would like to emphasize that a new face of the world industrial development, the universality of the major trends in technological progress serving both the purposes of the countries' national development, notwithstanding the level of their prosperity, and the global development raises interdependencies in economics and politics to an essentially new level and requires an urgent establishment of strong international structures and mechanisms.

High-Tech Development, Competition, and Industrial Culture in the Age of the Global Economy

Marc Giget

Technological Development at the Intersection of Competition and Cooperation between Companies

Today's world-wide technological revolution directly affects the business activity of industrial companies that are called upon to define and implement strategies. In addition to the follow-up and integration of new techniques specific to the field in which they operate, companies must position themselves with respect to accelerated developments in non-specific 'exogenous' technologies that have a significant impact on them.

The interaction of technologies undergoing rapid change thrusts companies into an environment characterized by uncertainty and growth in new opportunities. Traditional companies often feel that the devalorizing effect of new technologies on existing assets and expertise outweighs their valorizing effect, leading to low levels of investment, and consequently, to prolonged recession. On the contrary, emerging technologies (including microelectronics, biotechnologies, electron optics, and material manufacturing) establish directions for development among new companies, and for updating among older companies that are capable of integrating and valorizing those technologies.

Using Technology to Establish Identity

Current technological restructuring efforts are greatly altering factors that were previously decisive in corporate strategy. Radical technological transformations can quickly upset the positions of strength a firm has acquired on product lines or markets and thus render production infrastructure obsolete. Similarly, technological changes can seriously affect corporate financial strategies due to a rapid loss of value of traditionally profitable activities and the creation of a climate of uncertainty surrounding those that risk becoming unprofitable in the future.

Within this context, today's more dynamic industrial corporations are developing 'technological valorization' strategies, in which the company is the scene of two transformations: the search for all possible commercial applications of a series of technologies that the firm masters, and inversely, the search for possible technological answers to perceived needs in the markets it supplies.

Technological valorization strategies[1] have been visualised in a number of ways; all are variations on the same basic approach. We have most often used the tree to illustrate expertise based on an analysis of the strategic growth of several Japanese corporations[2]. The underlying feature of their strategy is the shift from a purely product/market strategy to one defined in terms of technological expertise. Technological expertise can be found at the centre of the very definition of the firm, whereas product lines and markets are peripheral. See enclosed Fig. 1.

The economic adage that says 'what is scarce is expensive' is familiar to all, even the man on the street. Companies have always organised their corporate strategy around the valorisation of a 'scarce factor' that must be used to best advantage. At certain points in history that scarce factor was raw materials, and strategies aimed to achieve vertical combinations in order to control the supply of raw materials. At other times it was access to markets, and strategies reflected the aim of controlling markets and distribution networks. At still other times, it has been labour and capital, and in each case, strategies were formulated for the purpose of gathering them, controlling them, and making them profitable.

Has technology become a scarce factor, prompting companies to place it at the heart of their strategy? That could appear odd amid today's veritable unleashing of technological innovations, for in theory, there is oversupply rather then shortage. Yet it is precisely this unleashing that renders technological investment highly vulnerable. What assurance does a company have, when it develops a anew microcomputer at high cost over several years, that that computer will not be outmoded by another, more powerful unit, even before it reaches the shelf and that its entire investment will not have been lost? Here, it not technology that is the scarce factor, but rather the technological lead over other companies, which must be protected and valorized as quickly as possible. The only way to hold that advantage durably is to remain at the leading edge of knowledge.

Technology as the Choice Area of Cooperation Between Companies

A technological lead is a tenuous advantage that firms must protect and valorize as quickly as possible. It can be considered a scarce factor in the economic sense of the term. Investment in R&D is very risky due to the continuous updating of interacting technologies, and few firms are capable of regulating the market in terms of technological generations. Mastery of technological development and innovation have become decisive factors in strategic competition between companies. Both have steadily moved to the fore among corporate managers' concerns over competition. By repositioning in terms of mastery of knowledge, companies that have launched technological valorization strategies have discovered - or more precisely, rediscovered - the veritable scarce factor.

For the very reason of firms' frequent inability to master successive technological generations, and due to high development costs, technological development has become the ideal point of cooperation between companies. There is no counting the scientific and technical cooperation agreements signed between major industrial groups, in all their forms. Some involve all the companies of a given country that are interested in a given field of technology, as is the case in Japan, with the multiple cooperative research centres supported by the Key Technology Centre. Others occur as part of major multilateral or bilateral cooperative technological programs, such as Esprit and Eureka in Europe. The complex relationships between competition and cooperation are quickly spreading world-wide along with the swift globalization of competition and markets.

The Impact of Technological Development on the Updating of Industrial Structures

Today's largest industrial firms originated and expanded in the climate of dynamic technological activity characteristic of the second half of the 19th century. The new technologies of that age included electricity, fine chemicals, automobiles, photography, radio, telegraph, telephone, and thermomechanics. Every one of those technologies gave rise to a number of companies, each with a strategy of transforming those technologies into a wide range of mass-produced goods. More recently, innovations have occured not only by applying new technologies, but also by combining new technologies with one another and with existing technologies, for example, electricity with electromechanics, electrometallurgy, electrochemistry, electrothermy, and electrolysis.

In the 20th century, industry in many countries moved steadily toward sectors that were narrowly defined and regulated in terms of products and markets. Crystallization into rigid sectors has been prominent in Europe, where increasingly restrictive administrative and social regulations have left companies less strategic leeway than in the United States or Japan, countries where government intervention in the industrial sphere is not as strong or costly to companies.

Over the past 15 years, industrial structure has broken down under the influence of the new technological revolution that is leading to a redefinition of industrial specialization. Emerging technologies in interaction bear names such as microelectronics, data processing, biotechnology, electron optics, 'mecatronics', robotics, composites, ceramics, and amorphous materials. Not only do they contribute to the creation of new products, they also radically transform existing ones.

Longevity through Knowledge

Present-day developments in the industrial system are different from the ones that marked the 19th century. Through this revolution has fostered the creation of many new scientific companies, it has also affected the large industrial corporations founded in the wake of last century's technological revolution, most of which are redeploying on the basis of accumulated knowledge and the incorporation of new technologies, as their initial field of specialisation falls into disuse. The strategic redeployment of European industrial groups has generally occured more belatedly than in Japan for structural reasons indicated above, yet it has now become widespread and is taking place at a speed that astonishes specialists of industrial economics.

To illustrate, Siemens successfully switched over from electricity to electronics and telecommunications, and it no longer manufactures a single product it built 40 years ago. Péchiney has moved from aluminium only to materials engineering and world leadership in packaging. Daimler-Benz, Europe's leading industrial group, is following the same pattern. On the occasion of its splendid 1986 festivities marking the company's 100th anniversary, and at a time when it had just begun its redeployment following one century of nothing but car-making, a West German automobile industry expert hailed that transformation with apprehension: 'Virtually overnight, Daimler-Benz has moved from a single-culture to a multi-product multi-sector firm.' He also questioned the risks of such a change. Now, four years later, 31% of the group's business is not car-related.

Unlike the strategy of American conglomerates of the 1960s and 1970s, which was essentially financial and oriented toward diversification,technological coherence is a fundamental aspect of the process currently taking place in Europe, which is essentially one of redeployment. It is closer to the changes that took place in Japan's industrial structures, despite basic differences in organisation, essentially the absence of heavyweight groups compromising a multitude of companies with interrelated activities.

By positioning themselves in terms of knowledge, science, and technological updating, the leading corporations contribute to their own longevity. It is striking to note in Europe that today's undeniably dynamic large corporations are at least one hundred years old. Daimler-Benz is 104, Nestle-124, the large West German chemicals companies BASF, Bayer, and Hoechst are 127 and 128, Rhône-Poulenc is 133, Siemens - 143, and Saint-Gobian - 325!

Even if the pace of innovation slows, as has been the case after previous technological revolutions, it is doubtful that the most dynamic companies will quickly abandon their special relationship to knowledge, science, and technology, which has ensured their longevity through times of fluctuating markets and product updating.

The Positive Macro-economic Effect of Technological Valorization Strategies

The continuous application of technological advances to satisfy demand in various sectors, which is characteristic of technological valorisation strategies, is a factor that supports overall growth. It runs counter to strategies of narrow sectoral segmentation, which subdivide the production apparatus between sectors in crisis and those that are profitable. Competitive strategies in crisis sectors are very costly to society (the case of steel and shipbuilding in Europe in the late 1970s and early 1980s), causing heavy taxation of profitable sectors and a diversion of investment.

Large inter-sectoral firms with technological valorisation strategies provide a highly effective means of transferring expertise and staff from crisis sectors to profitable ones, and of revitalizing activities on the decline. In this respect, Japan's major corporations have well demonstrated the efficiency of internal management of human resources and capital, shifting them from dying to growing sectors via mastery of a fields of technology.

These strategies also have a multiplier effect on the effectiveness of private-sector as well as government R&D expenditure. As a result of a lack of correlation between the breakdown of government credits for industrial R&D, which are extremely concentrated in specific fields (primarily defence, and aerospace) and the economic importance of the various segments of end demand, the redistributing role of technical advances played by corporations with a technological valorisation strategy is indispensable.

National Technology Policies and the Rise of Interdependance

The awakening of politicians to the growing importance of the technological factor in international economic competition was particularly strong in the early 1970's. Yet in Europe, and even more so in the United States, this awareness initially was very much political and little integrated into industrial culture. In 1974, Henry Kissenger told the American scientific community that the ability of the United States to exercise monetary control and govern the world in the style of the 1950s and 1960s had come to an end, and that the vehicle for action was the country's technological capability[3]. The myth thus developed that the rise in technology could be dematerialized.

The Drive for a Technological Edge in International Trade

In the 1970s many European and American politicians responsible for foreign trade predicted a growth in the share of 'gray matter' in international trade with respect to that of goods. Thomas A. Callaghan, chairman of Exim Bank, commenting on the spectacular recovery in 1974-75, following three years of deficit, said that technology seemed to be the magic word in international trade. Moreover, he said that markets closed to products are always open to technology, and that so long as the United States remains the dominant technological power in the world, markets closed to products will always be open to American technology[4].

Yet the share of dematerialized technology (patents, licenses, technical assistance) in international trade has virtually stagnated. It has only become a decisive competitive factor to the extent that it has been integrated into products.

The change in the American situation is exemplary: its extremely positive balance in terms of patents, licences, and technical assistance contracts (see enclosed figure 2) is indeed far from offsetting the general slump in the US trade balance, with the deficit widening to over USD 100 billion in 1985 and USD 150 billion in 1987. In the meantime, the Reagan administration committed itself to rebuilding American technological leadership and spent tens of billions of dollars on military R&D.

What has been the result of this monumental effort, and what is the explanation for the cleavage between technology and economic efficiency in the American industrial system? The answer to that question is not simple, though it appears that the Pentagon was the sole actor in the technological effort, that it was financed by the budget deficit, and that the beneficiaries were a handful of major corporations working almost exclusively in the US military market. There was a large gap separating those companies, sated with R&D credits and operating in the protected environment of preparations for a mythical 'Star Wars', and the rest of American industry, which was busy fighting a real trade war. Whereas military industry advanced in narrow fields of technology oriented toward very specific military applications[5], American industry on the whole began to lag behind in generic and distributive technologies that are applicable to a number of products. The high interest rates required to finance the growing budget deficit discouraged industrial investment, most notably R&D investment, as profitability was uncertain and in all cases distant.

Government intervention in technological development bears the stamp of specificity that differentiates the United States, Europe, and Japan, and it is far from being limited to the ratio of military to civilian R&D expenditure. The differences are fundamental (a detailed analysis would extend beyond the scope of this brief presentation). They involve a number of aspects, primarily:

- the ratio of public-to private sector funding for R&D (the consequences are very important, as the economic nature of R&D differs : as private-sector expenditure it is considered as investment, and as government expenditure it is seen it terms of markets)

- choice of orientation of government support toward fundamental research or technological demonstration

- support for training efforts intended to achieve control over technological changes.

From 'not invented here' to 'not invited here': the Risk of Technological Competition

The rapid globalization of the economy and the importance of technology in international competition and cooperation raise the issue of the relative efficiency of various governments' attempts to influence technological changes. The scope of this complex issue extends beyond industry to touch upon matters of defence, education, and culture. The strong cultural content of technology often leads western countries to develop the 'not invented here' reflex.

There is a risk that international technological relations could be perceived as relations of competition that could partially supplant political and military competition in intergovernmental relations, as was long the case between the United States and the Soviet Union. The determining role of mastery of new technologies has fostered technological protectionism in recent years, with countries turning inward and exhibiting exaggerated secrecy, resembling the 'not invented here' reflex.

Technology as the Ideal Crossroads of Future International Cooperation

There have been signs of harmonization of national science and technology policies, which can be seen primarily in the decreasing emphasis placed on military R&D, as the military use of R&D is increasingly being considered as an application of technologies that are generic by nature. They can also be seen in the development of precompetitive cooperative research with government support.

Moreover, there has been tangible growth in international scientific and technological cooperation, particularly in a series of areas where the efforts required exceed the possibilities of individual countries. This is the case for the industrialization of space, medical research, nuclear energy, and the airplanes of the future, which are gradually laying the foundation for increased interdependence.

Notes

[1] See 'Grappes technologiques', les nouvelles stratégies d'entreprises GEST, McGraw Hill 1986 (GEST is a group of researchers from LAREA, the CNRS research centre, and Euroconsult).

[2] See *The Bonzai trees of Japanese Industry*, Futures April 1988

[3] *Kissinger on science making the linkage with diplomacy,* Science 4, May 1974

[4] *European economic cooperation in military and civil technology*, Centre for Strategic and International Studies, Georgetown University, Washington Dc, September 1975.

[5] Such as very high energy lasers, magnetic propulsion of very high speed projectiles, and electronic defensive measures.

From Product/Market Strategy to Technological Strategy

Technologies as the means for creating the product/market lines charasteristic of the company

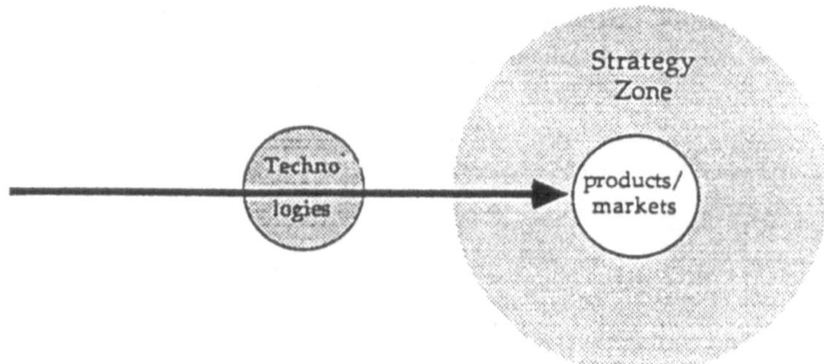

Versatile technologies as fields of expertise, characteristics of the company,

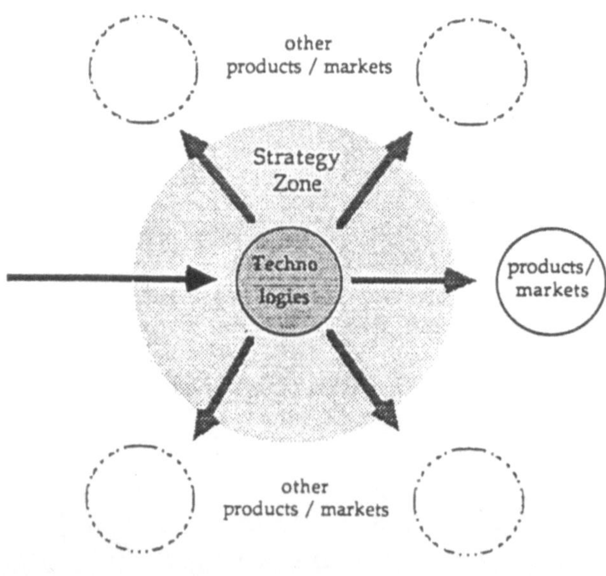

Euroconsult *FIGURE 1*

Trend in technological* balance of payments in major countries - 1970-1987

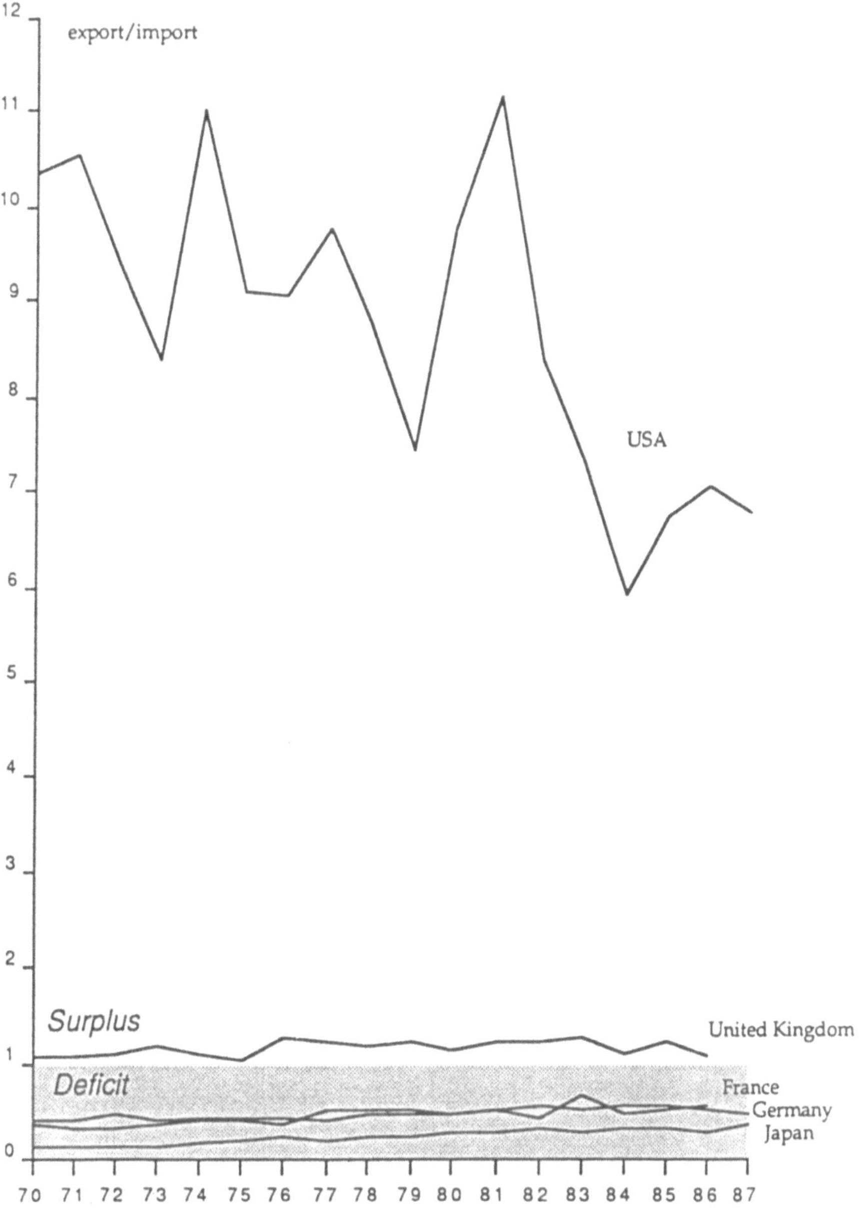

* patents, licences and technical assistance

Source of data : White Paper on Science & Technology *FIGURE 2*

New Trends in the Producer-Consumer Relationship: A Vital Aspect of Industrial Cultures

András Hernádi

Introduction

Processes of changes in the world economy have most often been examined from the viewpoint of production or foreign trade. However, a consumption-oriented approach would also appear proper on both logical and historical grounds.

Except for the so-called shortage-economies, it is rather consumption that determines production than the other way round. Most of the products and services appear in the market as a consequence of our earlier consumption, and only a comparatively small fraction of them enter it as new ones. But even if we take the case of our ancestor, homo sapiens did not start production by anticipating his later needs of consumption, but based on his previous consumer experiences. And likewise, it was only his instinct of self preservation that motivated him to consume for the very first time. But when he got down to production, and also during the process of production itself, he must have considered the joys of his earlier and future consumption. Finally, in the course of the historical development of mankind, production surpassing our own or our family's needs, ie. Surplus production for exchange was not started as a result of an ex ante decision. Rather as an ex post decision, based on the experience that our fellow creatures indicated an interest in the things we have produced, and showed readiness to exchange them for something else which they produced.

In his book, 'The Third Wave', Alvin Toffler goes on to say that it was only the industrial revolution that separated production and consumption from each other. Until then the role of exchange (or trade) was very limited. By the term ' third wave' he tries to replace or perhaps summarize ones like information age or electronic era, global village or post-industrial society etc., in which one can witness the gradual reunification of production and consumption, giving way to 'prosumption', and changing producers and consumers into 'prosumers' again.

In my view, it was the so-called do-it-yourself 'movement' that led to the reappearance of the 'prosumer'. A number of economic, psychological and social reasons can be found in the background, and the consequences will also be manifold for industrial cultures in the East and West alike. A detailed analysis of background factors and the most likely outcomes will be given in the paper.

Both in foreign and Hungarian economic literature, the analysis of changes in world economy has mostly been made by either a production or a foreign trade-oriented approach. It is not only in order to counterbalance these approaches that the present paper tries to put consumption in its focus. There are a number of historical and logical reasons to serve as a sound basis for doing so.

First of all, in an ever increasingly part of the world economy it is the so-called market mechanism that prevails. This way only the so-called shortage (or planned) economies would continue to be characterized by the priority of production over consumption. Should one say, however, that nothing can be consumed unless it has been produced before, I would argue , yes, that is true, but when living in our worlds today, we do not start from scratch. A great majority of the goods and services are already there in the market. Most of them are there because their producers had noted real or latent demand for them from our earlier consumption. When we consume something we simply cast our vote for its reproduction in smaller, identical, or bigger quantity. Even in the case of 'new' products and services, a number of marketing and advertising techniques are being used to generate, and then to call forth, a demand for them..

But even if we go back in history to the beginning, we would find our forebears *homo sapiens* not to have started production by anticipating later demands for consumption, but by making use of his earlier experience, ie. how well he could appease his hunger by eating the food he found as given by nature. In short, he made the decision to grow something that grew by itself before. And very similarly we might also say that our ancestors were not motivated in their consumption by anything else than their instinct of self-preservation or perhaps their pleasure, whereas when they got down to production, and also during the process of production, they must have thought of the joys of future consumption. Neither should one think that exchange-oriented production, replacing or supplementing the earlier subsistence farming, was the result of an ex-ante decision. Rather, it was started ex post, based on the experience that the other fellow (family or tribe) showed an extra demand for the food and was in the position of offering something else for it. Such 'deals', although at the beginning appearing to be of single and spontaneous character only, later became more frequent and deliberate, and led to changing the consumption-production-consumption cycle seemingly into a production-consumption-production cycle.

Let me hope, you will not take it as if I were merely were talking to myself if I touch upon a very vulgar, but both rational and historical, counter-argument, too. It is the well-known chicken-and-egg paradox that, according to public belief, cannot be solved. On the analogy of this paradox, however, we could even declare the relationship between production and consumption (one of) the chicken-and-egg dilemma(s) of economics. Without taking the risk of entering the field of theology or, for that matter, teleology as well, let me refer to the distinction in one of the above key sentences, where the phrase 'grew by itself' was confronted with the one 'he grew'. In my view, it is exactly this seemingly slight difference that can illustrate how the consumer of goods originally available as natural endowments has become a conscious producer of them in the course of historical development. In other words, apart from man's rise onto two feet and his usage of tools, one should attach special importance to his foresight as well.

What Does the Theory of Modern Economics Say on this Issue?

Samuelson and Nordhaus in their textbook 'Economics' (McGraw-Hill,1985) point out at the outset that market economies are ruled by consumers and technology, because the question of 'what?' (ie. what kind of goods should be produced) is answered by the consumers, even if 'an economy's resources, and the limitations of technology for transforming them into consumables, place limits on the consumer' (P.44). Similarly, when discussing the theory of supply and demand, they consider consumers to be like voters who use their money votes to buy what they want, 'and the people with the most dollar votes end up with the most influence over what gets produced' (p.59). This way they start the explanation of supply and demand from the side of demand, and it is only following the demand schedule and the demand curve that they analyse the supply schedule and the supply curve. For me it is hardly understandable why the English-speaking world refers to the relationship between supply and demand, using the words in most cases in this order, if they take the latter as the decisive one[1]. It cannot be just by chance that the same happens when they discuss aggregate supply and demand (pp.91-95). Even in the case of measuring national output, their first method for this is 'to add together all the consumption dollars spent for (these) final goods' (p.105), which is then followed by an earnings or cost approach (p.106).

Samuelson and Nordhaus do not stop at the conventional concept of GNP, but introduce the term NEW 'net economic welfare'[2]. When doing so they say, and this is a very important statement from the point of view of this paper:

> *'Suppose you decide, as you become more affluent, to work fewer hours, to get your psychic satisfactions from leisure as well as from goods and services. Then the measured GNP goes down even though welfare goes up... Consider also do-it-yourself work done in the home.... Because the values added are not bought or sold in the markets, they never enter into the goods and services of the GNP...' (pp.117-118)*

And a little later they add:

> *'Along with adding in 'goods' (e.g.,pleasurable air conditioning), GNP should be adjusted so that it subtracts out 'bads' (e.g., the pollution of air and water... and health risks that arise when coal is burned to generate electricity for the air conditioning).' (p.118)*

In the next chapter, 'Consumption and Investment', they show how the level of total output is determined by the interplay of spending and saving. In my own words, this is to illustrate how current and (directly or indirectly) deferred consumption determines the level of output. In my view, even investments can be considered actions of consumption (and not those of production), since by buying real estate, raw materials, machinery or labour we do not produce anything. It is only following these purchases that production can be started. Therefore, an investment is nothing else than giving up a part of today's consumption in the interest of increased or qualitatively different future consumption. As a matter of fact, Samuelson and Nordhaus also say that 'saving is the greatest luxury good of all' (p.125), and among the forces lying behind investment decisions they mention first of all 'the state of demand for the output produced by the new investment' (p.134).

Perhaps the best proof of priority given to demand-and therefore consumption as well-is the fact that when discussing supply-side economics, their analysis ends with the statement that in the USA

> *'... the torrential pace of economic activity in 1983-1984 was an expansion, fueled by demand-side growth, in the name of supply-side economics.' (p.192)*

If one goes through the whole book of 'Economics', it is always demand that precedes supply, and only after chapter 20 (p.433), can we read about business organisation, costs, competitive supply, while the theory of production itself comes up not earlier than in chapter 26 (p.579).

From Producers and Consumers - Prosumers again

Having talked far too much on the somewhat academic issue of the sequence of, or relationship between, production and consumption, it will hopefully be somewhat refreshing to touch upon a school of thought based on the realities of today's world and look ahead into the future. Alvin Toffler, in his book 'The Third Wave' (Bantam Books, 1981), says that following the 'invention' of agriculture approximately ten thousand years ago (called by him the first wave), it was only the industrial revolution (named the second wave) that separated production and consumption from each other. For until then the great majority of the products and services produced were consumed by the producers and their family members or of elites lucky and strong enough to get their hands on the surplus. Toffler goes on to say that

> *'... in most agricultural societies the great majority of people were peasants who huddled together in small, semi-isolated communities. They lived on a subsistence diet, growing just barely enough to keep themselves alive and their masters happy. Lacking the means for storing food over long periods, lacking the roads necessary to transport their product to distant markets, and well aware that any increase in output was likely to be confiscated by the slave-owner or feudal lord, they also lacked any great incentive to improve technology or increase production." (pp.37-38)*

It is true, and Toffler admits it too, that commerce did play a role even in these early times, but its importance compared to that of immediate self-use has remained marginal at least until the sixteenth century when, according to figures based on the research of the famous French economic historian, Fernand Braudel '60 percent or perhaps 70 percent of the overall production of the Mediterranean never entered the market economy.' Production and consumption were so close to each other that the Greeks and the Romans did not differentiate the two; they did not even have a word for consumer.

Following the industrial revolution, however, the so far characteristically subsistence farmers and producers started to reorient their activities towards the purpose of exchange. Very likely this was the first time mankind faced the idea of market orientation. Man ceased to manifest the unity of production and consumption, because less and less people were consuming only the things they produced by themselves.

Toffler introduced the term 'the third wave' in order to replace or perhaps to combine the ones which he was not satisfied with, like 'space age', 'information age', 'electronic era', 'global village', 'post-industrial society' or 'scientific-technological revolution'. The 'third wave' brings with itself the return to the unity of producer and consumer in the (re)-emergence of the 'prosumer'. He is, of course, aware of the fact that certain unpaid activities, e.g. the ones called non-economic in market economies, like household work or bringing up children, have always been a part of the sector of self-sufficiency. I would hastily add, however, that as a part of the marketization process accompanying the industrial revolution a number of such activities also appeared as objects (or rather subjects) of exchange: rich people could always afford to hire wet nurses, servants, private tutors etc.

In my view, the greatest change in this respect has taken place by the appearance of 'do-it-yourself' initiatives. Although, originally it must have been just a tricky idea of merchandising that a certain, most often the final, phase of the production or servicing process could and should be left over to the consumer, by giving him the joy of active participation, it has opened up new vistas for sales too[3].

A number of psychological reasons were also in favour of letting the consumer *dot his i*: finish the process of assembling some furniture or household equipment, or cooking some canned or deep-frozen food (The famous Swedish furniture manufacturer IKEA, for example, sells most of its products with the slogan 'You can take it with you. At home it all comes together'). The increasing demand for services all over the world also resulted in technological and marketing techniques, which led to a modification in the division of labour between the producer (or the seller) and the consumer (or the buyer). This is how the operator-controlled telephone calls were replaced by direct dialling; self-service was introduced in department stores, restaurants, banks and gas stations; and even the task of registering in a hotel requires guests to fill in a registration form.

In all these cases, supposing there has really been some price-cut as a kind of compensation for the tasks taken over, the buyer has 'paid himself' and thus from consumer he has partly become producer as well. Nota bene, he could earn some tax-free income too...

I believe it is worth mentioning that some further elements of this process appeared both constraints and possibilities. Perhaps the most important one among them is that at a time of rising prices producers and consumers alike are interested in eliminating all possible factors of cost. Therefore the former, following some thorough market investigations and suitable technological developments, would come out with products that need some after-purchase do-it-yourself activity. A special reference is due to those savings and cost-reducing methods that are in favour of 'thinking' in units, as opposed to packing, storing and transporting the goods in a pre-assembled form. Owing to reduced prices the number of potential buyers would increase. But apart from these financial aspects one should also bear in mind some other trends too. For example, the fact that nowadays more and more people (have to) live in so-called housing estates, which would normally have smaller apartments and rooms, makes 'transporting-in-one-piece' almost impossible. Do-it-yourself assembling, at least in planned economies, became popular because of bad experiences too: we came to realize that something had been done properly only if we ourselves had done it. Chatting with, feeding, and giving drinks to assembly-men (let alone repairmen) in one's home has proved to be a nuisance worth avoiding at almost any cost.

The popularity of do-it-yourself activities has obviously been promoted by industry itself. An arsenal of handtools and accessories has been developed and made available in shops. We might even say that a new branch of industry was started. Parallel with this, education and self-education followed suit: with the exception of learning languages, it was technical education that came to the fore thus ensuring the necessary mental and practical capabilities for changing consumers partly into producers again. The same process has been supported by a great number on 'How to...' or 'All you have to know about...' publications, not to mention technical books. It is, of course, also true that these forms of (self)-education were not without antecedents: we should not forget that all of us had our childhood experience of playing with building blocks, kneading and modelling clay, jigsaw puzzles, Marklin, Schuco, Duplo, and Lego toys. From the point of view of our theme, the most important lesson we can draw from this is that all of us have gone through the phase when, at least once, we were both the producers and consumers of our own fantasies and skills.

While in the planned economies it was mostly poverty or more recently pauperization, and the poor quality of goods and services, the shortage of spare parts and components that called for do-it-yourself, the process of alienation has also played a role in both systems in motivating people towards these activities. Basically it is still true that in the poorer countries of Eastern Europe do-it-yourself should be considered as a kind of necessary extra work, whereas in the richer Western societies it is more of a hobby, something that you do in your free time, in order to achieve better self-realization. I believe it is worth mentioning here that, quite often, it is this latter chance that makes it possible to change one's career in the West. In Eastern Europe, however, it is rather 'failure stories', as opposed to success stories, which follow from the over planned character of education and training, the well-known counter selective mechanisms having prevailed for at least four decades. As well, the long oppressed entrepreneurial spirit, and the immobility of labour originating from an almost notorious housing shortage were the typical systemic reasons that were to blame in the Eastern European countries.

Likely and Desirable Consequences

All this, however, brings us to the issue of new work and life-styles of the reborn prosumer. According to Toffler, we are heading for an economy in which many people will never have a full-time job, and the usual differentiation of leisure and work will be replaced by that of unpaid and paid work.

> *'Given home computers, given seeds genetically designed for urban or even apartment agriculture, given cheap home tools for working plastic, given new*

materials, adhesives, and membranes, and given free technical advice available over the telephone lines, with instructions perhaps flickering on the TV or computer screen, it becomes possible to create life-styles that are more rounded and varied, less monotonous, more creatively satisfying, and less market-intensive than those that typified Second Wave civilisation' (pp.278-279)[4]

The most striking element of this new life-style will be a radical change in the concept and role of 'home'. Parallel with the well-known extension of service sector activities, an ever increasing part of the remaining agricultural and industrial work, so far of physical character, will become intellectualized, thus becoming unbound from the site of production and services (ie. the farm, the factory, the office etc.). Work done from home, which is definitely on the increase in our time, can, in fact, bring about important changes. Let us consider the social aspects of women entering the labour force, the crises of the family, and the economic gains related to savings on investment and operating costs. A tremendous amount of savings on time, fuel, and pollution become possible by moving information electronically instead of transporting masses of people from home to work (and back home) every day[5].

All these epoch-making changes, which we are not only witnessing but more and more participating in, also show a direction where man will bring the world surrounding him and mostly shaped by him, too, under his control again. Thanks to automation, computer aided design and manufacturing (CAD and CAM), he will return from the alienating monotony of mass production to the areas of self-liberating, creative activities. At the same time, by making use of the advantages offered by CAD and CAM, his individual needs will be incorporated into the process of mass production. He will make his participation in mammoth organisations bearable by cutting them from inside into smaller units or groups, or by placing himself outside of them (i.e. into his own home) he could alleviate, if not put an end to, his unnatural relationship with them. He will as well reconsider the value judgements imprinted in him, but proved to be false in the course of previous centuries. He will change his wasteful and conspicuous consumption patterns into thrifty and rational ones, replace the production processes and life-styles that are equally harmful to the ecosystem and to himself by environmentally and human friendly ones. By realizing that to have and consume goods is not the ultimate goal but just one of the means of human welfare, he will hopefully give up the traditional pattern of consumption or profit-maximizing and accept the optimizing attitude of alternative economics.

Notes

1 This is one of those rare cases when the Hungarian usage of words seems to be more precise, since we always speak about the relation between demand and supply, in this order. Although it is outside the framework of the present paper, I still find it worth while citing the well-known example of differences among various languages in expressing the same terms: while people in the English-speaking world 'earn' money, the French gain it

(gagner l'argent), the Germans serve for it (Geld verdienen), the Russians work for it (den'gi zarabatyvat'), and we Hungarians look for it (penzt keres).

[2] The authors refer to a study by Nordhaus, W & Tobin, J (1979) *Is growth Obsolete?* Fiftieth Anniversary Colloquium V, National Bureau of Economic Research, Columbia University Press, New York, and point out that the concept of GNP includes many elements of ambiguous contribution to our individual welfare, while it excludes a number of important ones. Op. cit. pp.117-119.

[3] In 1954 the value of goods sold for do-it-yourself purposes in the USA was estimated at 6 billion dollars, 2.5 per cent of all consumer expenditures. Since the value of do-it-yourself work itself has on average been two or three times that of the goods used up, all in all, this kind of activity could well represent some 10 per cent of all consumer expenditures. US Department of Commerce, Small Business Administration, 'Summary of Information on the Do-it-Yourself Market', Business Service Bulletin, No 84, November 1954. Cited in T.Scitovsky: The Joyless Economy. Oxford University Press, 1976,pp.279-280.

[4] This vision of the future has since been depicted by some other authors of the school of alternative economics, too, as one of the possible alternatives of the post-full-employment era. J. Robertson, for example, writes about the possibility of 'ownwork', apart from 'wide-ranging' employment and 'leisure'. In Ekins, E P (ed) (1986) *The Living Economy* Routledge and Kegan Paul, pp 85-96.

[5] More details on this latter issue have been mentioned in my 'A hazai automobilizmusrol' (Automobilism in Hungary), *Valosag*, April 1989, pp.48-58.

Impacts of Information Technology on Japanese Industrial Culture

Takao Nuki

Introduction

Micro-electronics Technology which is the basis of computerisation or automation of factory (FA), office (OA), store (SA) and home (HA) are having some impacts on our traditional cultures. We define technology and culture (and their roles) as below:

Technology = 'intellectual factor of method' or *'a set of knowledge about the methods of achieving given goals'*

- its role = 'promotion of our material abilities (for production, communication, medical care, military affairs, etc.) to fulfil our desires'

Culture = 'a set of social values and senses' or *'the pattern of behaviour which is based upon a set of social values and senses'*

- its role = 'formation of a social order without violence and war' or *'control of mankind's material ability i.e. social decision of technology goals'.*

So, technology is the engine and culture is the brake and handle to our life. But culture itself will be deeply influenced by the contemporary technological innovation. And, as a result, the gap between technology and culture will spread further. Especially this is likely in Oriental (including Japanese) culture because of the Western-originated methodology of modern science and technology. The purpose of this paper is to consider the cause of this spreading gap and its effects on Japanese culture focusing on the effect of information technology.

As for biotechnology we can imagine the conflict between technological possibilities and our culture more easily than the conflict caused by information technology (micro-electronics). In the near future we may have, for example, the possibility of choice to decide whether we must accept biotechnology to create a new species of feather-less chicken for the purpose of saving time and labour of plucking feathers before shipping the broiler. Though the effect of micro-electronics on our cultures is not so clear as the effect of biotechnology, we should also be aware of the potential threat of micro-electronics to our culture. It is our culture which decides the goal of technology finally. But it is the logic of economy that decides the goal of technology directly, because technology is developed or adopted mainly by economic bodies (the enterprises) which do not necessarily regard the culture as of the first priority. No doubt that the enterprises cannot ignore the culture so long as it affects the market of their products. Nontheless, the power balance between economy and culture does not always assure that the culture carries through its logic.

Accumulation Capacity of Technology and Accumulation Incapacity of Culture - The Gap in Levels between Technology and Culture

Computerisation means the development of technology for information processing, storage, retrieval and transmission. The development of technology is now expressed as the development of hardware like machinery or apparatus which show improved performances, and also as the development of software.

Both for the hardware or software, the essence of technology is knowledge about the method to achieve the given goals. For example, a machine is a composition of some factors which include material (of the machine), money (which was necessary to produce it), labour (which was done to produce it) and knowledge. Technology as the intellectual factor of a machine is the knowledge about the structure, the quality of the material, the process of manufacturing, etc. of the machine. Material, money and labour are necessary factors for giving concrete shape of the knowledge.

Knowledge means information as long as it can be transmitted in the form of a blueprint, a sentence, data or a real sample. And because it is information, we can transfer it from person to person, from generation to generation and, by adding new improvements or inventions, technology can accumulate with the passage of time. So, the development of information technology may contribute much and accelerate the development of the whole field of technology; For example, the automatic translating machine is already put to practical use in the domain of technical documents and commercial correspondence.

As a result of the accumulation capacity of technology, we have realised great progress of material ability. Boarding the Concorde supersonic transport, for example, we can move with a speed of about 550 times as fast as when we walk. Concerning technology, we start from the last developmental stage of the former generation. Thus even the present-day schoolboys may have more technological knowledge than first-class scientists or engineers of ancient Greece or the Roman empire.

But how about the level of human spirit or culture? In the light of the fact that human beings cannot erase war from the earth and thus they risk self-destruction by nuclear war, we cannot say that we obtained a higher level of human spirit or culture.

As already shown at the beginning of this paper, culture is a pattern of behaviour. We cannot attain it by learning only. It must be accepted at the deep level of mental structure, and this deep mental structure is not easily changeable. For the matter of spirit can we start from the last developmental stage of the former generation? We can learn about the achievement of Christ or Buddha. We know what they preached or what they did. But, with this knowledge we cannot reach the same dimension of spirit of Christ or Buddha. We can believe in them and make efforts to reach them. That is all. The field of spiritual enlightenment that was obtained by the long term individual cultivation of the mind may disappear completely after his death. There is no innovation and no accumulative development (cf. figure 1). The reason why we cannot simply be very glad of having attained technology development is that we know or sense that technology development may cause spreading the gap between the level of technology and culture and some day, human beings who remain at the lower level of culture may make a mistake in using their very strong material power.

If technology development will prepare the causes of ruin for human beings who are torchbearers of technology (and culture), the process of technology development is, at the same time, the process of destruction of technology (and culture).

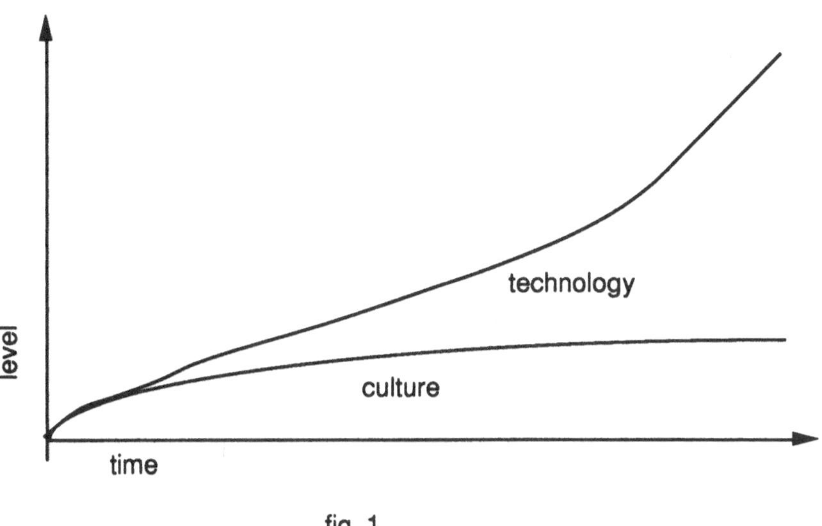

fig. 1

During the Gulf war information technology made a great contribution to planning and implementation of operation Desert Storm, whereas it had taken almost no important role to avoid the war. Information technology has improved the accuracy of attack against the enemy or the effectiveness of protection from the enemy. But as for the means to persuade, to negotiate or to agree, we cannot see any notable contribution of the technology simply because technology cannot necessarily promote the wisdom of human beings to solve conflict peacefully.

After the second world war the Japanese constitution forbids waging war and to send military force to foreign countries. And because of the hatred against the misery caused by the war the majority of Japanese have accepted and supported the constitution though it was established under the regime of American occupation. The Gulf war is now raising a serious argument about whether the Japanese government could send its military force to foreign countries when it is requested by the United Nations. Thus the Gulf war produced a turning point of the post-WWII history of Japanese peace policy which has been part of the basic framework of the economic growth of Japan after WWII.

Like other countries, military technology, i.e. the technology of destruction, in Japan has proceeded far more than that of the period before the second war largely owing to the development of micro-electronics (though Japan has banned the export of any weaponry). Who can assure that the culture or wisdom of Japan has proceeded far more than that of the period before the war to be mature enough to avoid another war?

The Gap of the Life-cycles between Technology, Culture and Human Existence

When we think that technological cycle means the length of the period from the emergence of a technology till its obsolescence because of new substitutional or major technology, it is very clear that technological cycle has been shortening while also accelerating.

The cultural cycle which means the life-span of a culture also has been considerably shortened by the influence of the technological cycle. Because, even though technology cannot have much influences on the *level* of culture, it can have influences upon the *type* of culture. For example, computerisation is providing more opportunity for the jobs which are suitable to women and this is changing the social perception of the women's role as well as the industrial culture in Japan.

In this century Japan has experienced three eras in which the majority of working people have changed from farmer to worker in manufacturing industries and then to worker in the service sector. The peasant (farmer) culture has been almost expelled and the culture of manufacturing worker is becoming minor.

On the contrary, the human life cycle, i.e. the length of human life has been much extended owing to the development of medical technology (ex. CT scanner, automatic diagnostic apparatus, etc.). In Japan, the average length of male life already exceeds 75 years, with about 79 years for female life expectation, and now they say 80 years of life. Consequently, the length of human life-cycle became longer than technological and cultural cycles. The technological cycle is too fast for adapting sufficiently to culture and also the cultural cycle is too fast for the middle or older generations.

Culture is bewildered by technological changes and people are bewildered by both technological and cultural changes. In earlier times there was no remarkable change of technology and culture in a human life. Now, in a human life, culture changes many times and technology changes more frequently. We can show this fact by the figure at next page.

Earlier days

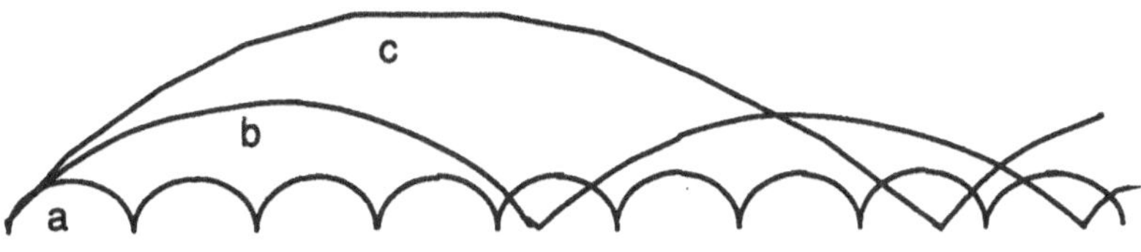

a : human life cycle
b : technological cycle
c : cultural cycle

Present days

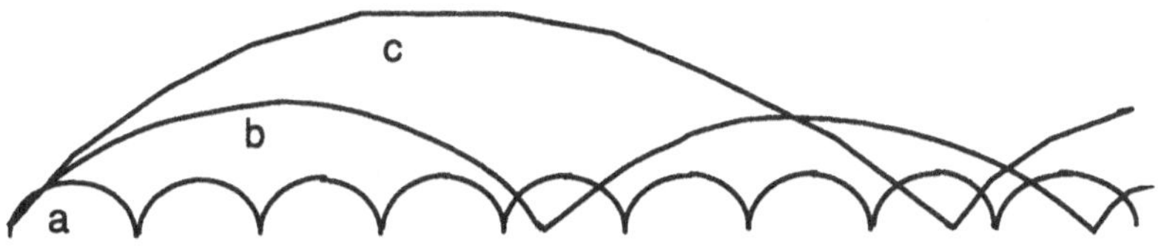

a : technological cycle
b : cultural cycle
c : human life cycle

fig. 2

What are the consequences of this shortening of technological cycle? Younger generations can adapt to new technology more easily than older generations. The level of skill can be higher in proportion to the years of experience, but older age may become a barrier to understanding and adapting to new technologies. That means the breakdown of a proportional relation between the ages and work ability. Thus it will influence seriously the social order between the older and younger generations which has been an important part of our traditional Oriental morale. The research by the Japanese Ministry of Labour shows that more than fifty per cent of Japanese big companies feel the over-employment of employees who are more than 45 years old. The reasons are shown in figure 3.

The reasons why the companies regard the middle and older-age emloyees as " surplus "

Lack of managerial position which fit their age

They are a barrier to utilising the power of younger generation

The cost of salaries are too expensive in comparison with other generations

The contribution to the company has leveled off in comparison with the level of salaries

They cannot adapt to new technologies or marketing

The adjustment opportunity in the form of "dispatching" has reached its limit

They cannot adapt to the changes of managerial strategy like finding new business domains or develop production in foreign countries

others

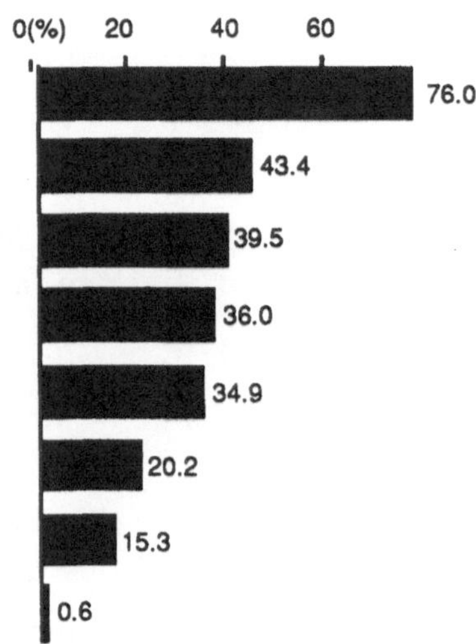

fig. 3

'Transfer of relation' - From Man-System to Machine-System (A Zero-sum Relation between Technology and Culture)

It was already pointed out that by the development of technology the human skill of the skilled worker is transferred to the improved machine and the types of work are decomposed into complex and simple work. But it must be also pointed out that development of technology in the form of computerisation of work implies the transfer of human relations from man-system to machine-system or computer network system.

Through factory or office automation, the opportunities and necessity of direct communication between workers has been eliminated drastically and indirect communication through the mediation of computer and its terminal became the major part of contact. In short, man-machine interface has substituted for man-man interface. The worker works facing a computer terminal, not another person. Productivity of work may rise, but personal contacts or relations between workers are absorbed by the machine (computer network) system, and as a result the relations between workers are thinned-out to that extent. (cf. figures 4 and 5)

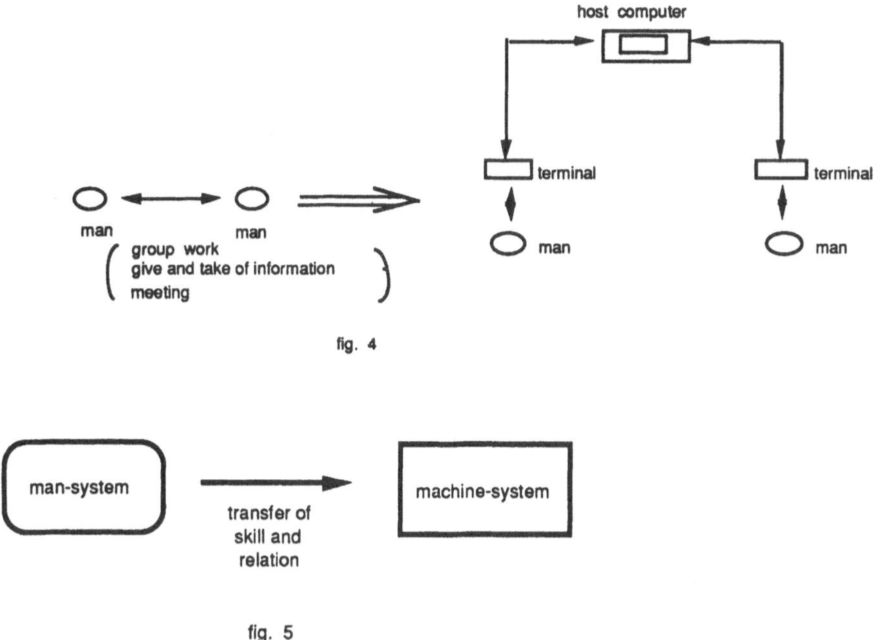

fig. 4

fig. 5

Saving the necessity of direct personal contact for doing their job has influences upon the mental relation like fellow feeling or sense of solidarity. In the given conditions, the degree of these feelings or this sense of solidarity is proportional to the frequency and directness of contact. So, this transfer of relation leads to isolation of worker and implies that the community spirit in the organisation will be weakened if other measures for strengthening this spirit are not brought in.

In Japanese enterprises the harmony and tacit understanding within members of an organisation which is based upon follow feeling, sense of solidarity and community spirit is the most important factor for doing a task effectively and for finding the working life worth living. Thus the weakening of group or community oriented values produced by 'transfer of relation' will be a blow to the traditional Japanese (Oriental) values. In general, Japan runs the risk of unexpected destruction of its traditional corporate culture without having prepared an alternative.

Furthermore, it is worth noting the other factor which promotes the isolation of workers in organisation i.e. distrust in the information society. We expect that the information society will create a society in which people can use or exchange more information and consequently understand and trust each other much more. Partly this is true. But the information society also means a society in which more information can circulate in the same way as economic goods; and one of the important features of information as an economic good is that there is a huge gap between the cost of creating information and the cost of replicating it - in the case of manufactured goods, manufacturing facilities, skills and raw materials are needed just to make one item - while the use-value of both is the same. Namely, the cost of making a copy of some information is almost negligible (only the cost of some sheets of paper or floppy disk). For this reason, in the information society, fear is often felt about the possibility of having information stolen. But to be cautious is, in other words, to be distrustful. Thus we recognize two contradictory effects of the information society on our mental being.

In private enterprise the sign of this distrust is already clear. There are many buildings, rooms, or zones to which workers of the enterprise itself are barred from entering without showing their identification cards. And again they are not allowed to access information without the code number. It almost seems as if we are entering a military area! Maybe there is no choice if we want to protect information. This is very similar to the management style of distrust where trust in their own workers doesn't exist. As a result of highly homogeneous society, our traditional attitude in Japan when we meet someone who is a stranger, has always been to treat him as an ally until he proves himself to be an enemy. Therefore one of our anxieties is that the impact of computerisation would change the traditional character of Japanese style of management, which has put confidence in employees. The information society would undermine the relationship of mutual trust which has been the main factor behind the competitive edge of Japanese industry.

Conclusion - Our task

With the restriction of natural resources and environmental pollution, it is now clear that we cannot expect the ideal society. That is so say, we would not be able to fulfil our desire maximally by the limitless production capacity which would be attained by great strides in technological development, whether in capitalist countries or in socialist countries. So, we - particularly the advanced countries - must check our values which regard the maximization of desire fulfilment as worthy. And we must create the view of Nature that Nature is not an object for conquest to bring out maximum fulfilment of our desire, but Nature is the partner of our living, Nature is the absolute condition for our survival. We must also create the values that consider a life with humble consumption of material goods and a lot of sympathy with other people as most worthy. Our civilization has been developed in the direction of affirming our desires and of supplying various measures to fulfil the desires. Now we are arriving at the point where our desires themselves should be reconsidered to create a new view of values. To create a new view of nature and values means to create a new culture. Technology will develop with passage of times because the logic of economy strongly requires new technology for economic growth. So, the maintenance and development of culture must be given the priority as the object of our conscious effort. - But the opinion above must not be put to use as an excuse for fixing the differentials of technology and living standard between industrialized countries and developing countries.

I said above that we must create a new culture. But we have this type of values already in traditional Oriental thought. Actually, Dr. Schumacher, who is the author of the book 'Small is Beautiful', placed high value on Oriental thought, especially on Buddhism, as the base to solve the problems caused by modern technology - exhaustion of resources, pollution of environment, unemployment, alienation.

But, originally, the Buddhist priest lived and did his training by asking charity to get his food. Then, Buddhism does not have the tradition or experience to take part positively in the production activity. Consequently, for example, in the series of question and answer for the training program of Zen, an important denomination of Buddhism, we cannot find any question concerning the problems of production or technology. The task of placing technology and culture in relation with nature and human beings is a difficult challenge also for Oriental thought.

Other papers related to this paper written in English by the author:

Nuki, T (1983) The effect of micro-electronics on the Japanese style of management. In Labour and Society Vol.8, No.4, October-December 1983.

Nuki, T (1990) The 'Transfer of Skill' and the 'Transfer of Human Relations' to Machine Systems. In AI & Society Vol.4, No.3 July-September 1990.

Japanese Human Resource Management in the Cross-Cultural Interface: A Case Study of US-Sited Japanese Subsidiaries

Kazuo Takeuchi

Introduction

As Japanese direct investment is increasing very rapidly, many Japanese companies build and purchase factories, and start operation in foreign countries. In foreign countries there exist more or less different values, life styles, and business practices. On the other hand, Japanese expatriates who are assigned to overseas operation have been inculcated with Japanese values, business practices, and work ways while they worked in Japan. When they start working in US-sited Japanese subsidiaries, two different industrial cultures begin to interface. If there is wide discrepancy between the two cultures, conflict, complaint, dissatisfaction, and anxiety about future employment will arise.

The purpose of this paper is to analyse Japanese human resource management in the cross-cultural interface in the US-sited Japanese subsidiaries, primarily based on case studies conducted during the period from 1980 through 1988, while the author visited the US as a visiting scholar with additional short visits for surveys. In the United States, people with a variety of cultural backgrounds and different values work in the same workplace. Often new immigrants with poor English proficiency work on the factory floor together with Americans. In the different work environment, how is Japanese human resource management modified? What practices of Japanese human resource management are accepted by American employees and what are rejected by them? These issues will be analysed and discussed in this paper.

The author is very grateful to the Japanese subsidiaries and American companies which welcomed his visits for interviews, although their names will be kept confidential.

The Features of Japanese Values Underlying Japanese Human Resource Management

First of all the features of Japanese values underlying Japanese human resource management practices will be defined.

Collectivism

Japan has been described as a country of collectivism in contrast to America, which is described as a country of individualism. Collectivism and individualism are psychological tendencies and are reflected in subtle behaviours. Both tendencies are present in both American and Japanese industry, but the usual and preferred tendency in Japan is collectivism.

Collectivism can be defined as the attitudes of people who view the goals, interest, or values of the groups or organisations to which they belong as more important than those of individual members. In Japanese society collectivism is pervasive. People tend to think of other's interests and to have their own behaviour conforming to the norms of the group to which they belong.

The process of reinforcement of collectivism begins in childhood. Families insist that children should think of the feelings and desires of siblings, other relatives, and friends. School teachers, including those of nursery schools, insist on collective behaviour and discipline. Children are taught to accept responsibility as a group. Collectivism is also reinforced in Japanese Universities by collegiate club activities. Unlike American Universities, the 'loner' or individual who refuses to join seminars or clubs is not respected for such behaviour.

Additionally the culmination of the reinforcement process of collectivism as a way of life in Japan takes place at work. Most organisations have rigourous indoctrination programs which are designed to make certain that all employees understand and accept organizational goals as being of paramount importance. Since opportunities to move from firm to firm are limited and opportunities within the firm depend so heavily on willingness to conform, the addition of this intense economic pressure to the already pervasive social pressure guarantees that collectivism will be the dominant attitude. American firms, after a shorter indoctrination and training period, adopt the attitude of telling the new employees to demonstrate what he or she can do as an individual.

Collectivism, as a basic ideology in Japan, strongly affects the various aspects of modern business practices in japan. In a collective organisation there is a great deal of mutual trust. When there is reduced fear that someone will claim individual credit or acclaim - the group gets it. While top management in the US expect some serious competition among company executives who are looking for individual rewards, such behaviour in Japan could destroy the foundation of the company, so it is not tolerated at any level.

Many of the large companies in Japan belong to enterprise groups, such as the Mitsubishi group, the Mitsui group or the Sumitomo group. Member companies of the group hold the stocks of other member companies, and also form the network of mutual transaction of products, parts and components, and also financial support. Additionally large companies form a group of subsidiaries and cooperative subcontractors, so called 'Keuretsu', which produce parts and components primarily for their parent company and also accept executive retirees from the parent company.

Japanese management know that collectivism functions best in small group activities. In addition to emphasising the collective attitudes at the smallest formal group level in the operational structure, Japanese management encourages many other small group activities. These small groups cross departmental working lives and serve as additional cores for the reinforcement of the collective ethic. There are study groups, sports groups, hobby groups as well as well-known quality circles in factories and offices. Thus the reinforcement of the collective ethic is a result of rational, positive action on the part of Japanese management, who feel that collectivism gives Japanese industry a differential advantage over other organizational structures.

Recently Japanese managers indicated that they feel it is more difficult to manage young employees than older employees. They complain about the more individualistic trend of the younger generations. In Japan, being individualistic implies being egoistic, and this is not in keeping with the collective ethic. The collectivism and work ethic in Japan have been changing, and labour turnover has been increasing because of recent labour shortage, but the Japanese management is confident that they will collectively adapt to the changing conditions (Berkstresser and Takeuchi, 1983).

Egalitarianism

Egalitarianism in Japan typically takes place in small social and economic disparity between high and low organisational levels. Egalitarianism combined with collectivism is embodied in short psychological distances and open communications between superiors and subordinates, and a high level of sharing of management information among employees. The concept of short psychological distance between managers and ordinary employees can be seen in the policies that Japanese top management, middle management, engineers and workers wear the same uniforms in the factory, eat in the same cafeteria, and also managers and office workers work together in the 'open-space' office without much prerogative for managers.

Much important financial and technological information is available to employees on a routine basis. Japanese managers and employees do attend many departmental and interdepartmental meetings, which is a form of sharing information. These frequent meetings do facilitate vertical, horizontal and oblique communication within the firm and reinforce the collective attitude of Japanese managers and employees. It has been said that Japan is the vertical society, and the vertical human relationship based on age and seniority characterize Japanese companies.

However, the openness of communications with this vertical structure - the willingness of managers to accept subordinate's ideas, opinions, and even criticisms of managerial decisions and policies - make it effectively less vertical than the much less open superior-subordinate structure found in most American companies (Berkstresser and Takeuchi, 1983).

Egalitarianism can also be found in the Japanese belief that there is not much disparity between individual potentials, and every individual can acquire a certain level of capability if they receive education and training and make efforts by themselves. Strong attitudes against the government policy to introduce the class system to classify and organize students at schools in accord with academic ability still do exist among parents and teachers.

Family-type Management

As shown in 'Kaisha-ism', which implies the strong tendency of Japanese people to identify themselves with the company which they work for, the company is the place where Japanese employees make friends, eat, drink and chatter with superiors, subordinates and peers. They also enjoy sports and leisure together, feel joys and sorrows and help each other in wedding ceremonies and funerals. At the risk of exaggeration, the whole lives of Japanese employees are strongly oriented to and are interwoven with the company in which they work. Therefore, if the peer employee should stay late at the office to complete his or her job, the other employees are supposed to stay late to help the peer's work voluntarily without the order from the superior. When the peer moves to a different place for relocation, all employees in the same department are supposed to go to the peer's home to help relocation on holidays, such as Saturday and Sunday.

Occasionally superiors make telephone calls to the young subordinates' apartment late at night to make sure he or she stays at home. The concept underlying the manager's extraordinary concern over the young employees' private lives is that the company is taking care of young employees on behalf of their parents. Young employees do not welcome the superiors' concern in much the same way as their parents' concern and are likely to be embarrassed rather than grateful.

The Features of Japanese Human Resource Management

Based on the Japanese values mentioned above, the following Japanese management practices are commonly used in Japan.

Hiring Young, New Graduates

In Japanese companies, hiring young, new graduates right after their graduation has been the most common practice. Recent labour shortage and changes of work ethic of employees, especially young employees as well as environmental changes such as globalization and technological innovation have resulted in the increasing labour turnover, and hiring experienced workers is becoming popular. However, the staffing practice to hire young, new graduates right after the graduation still is and will be the essential and common practice as the basic condition for the long term training and development of human resources and seniority-based compensation system with the low starting salaries of young employees in order to cut back total labour costs

Long Orientation for New Employees

Long orientation is practiced to indoctrinate new graduate employees with corporate culture as well as the functions and organizational structure of the company. In order to indoctrinate new employees with corporate culture, a company's history and corporate philosophy are taught.

Often the orientation courses include physical and mental training such as Zen, long-distance jogging, and bathing in the cold water of the river or ocean. Additionally, after new employees are placed, the indoctrination of corporate philosophy and culture through the actual job experiences in the workplace will continue.

Job rotation

In Japan, horizontal job rotations crossing the functional boundaries are practiced every two or three years except for the specialists of particular departments such as accounting, finance, and personnel in order for employees to gain wide job experience and views on the management of the company. Through frequent horizontal job rotations, employees acquire a good understanding of organizational structure, departmental functions, and establish wide human relations within the company, which will also be effective for promotion and lifelong employment.

Promotion from Within

Based on the training and development through frequent horizontal job rotations, promotion from within is the common practice in Japan. Staffing managerial positions directly from outside is exceptional except when no qualified candidates can be found within the firm in the case of fast diversification or global expansion. In the lifelong employment system, the failures of promoting employees from within will destroy the loyalty of employees and will result in low performance of the company.

The feature of Japanese personnel appraisal is that in promotion and upgrading employees' rankings, job attitudes such as willingness for cooperation, acceptance of responsibility, loyalty, spontaneous efforts for acquiring more maturity and skills, superior-subordinate relations as well as job performance and capability are evaluated. In addition, superiors are often much concerned about subordinates' private lives, which informally affects the superiors' judgement on the total evaluation of subordinates. In most Japanese companies it takes a long time for employees to reach the first managerial positions. Many large Japanese companies used to promote together all young college graduate employees hired in the same year without any delays for about ten years until they reach the first managerial positions.

Promotion decisions are based on seniority, attitude, capability, and performance, while in the US capability and performance are the main factors for promotion decisions. In qualification systems which are the basis of human resource management in Japan, the waiting period from three to five years is common in upgrading employees from a particular qualification level to the next. Even in high technology companies, promoting young employees to managerial positions over older employees is considered as an undesirable decision. The fact of a young talented manager managing older subordinates still scares people. Reversal of seniority in promotion which is frequently observed in American companies is a scary nightmare for most Japanese companies.

In Japan, compensation is composed of two different basic compensations, such as seniority-based and qualification-based compensation. 'Qualification' implies general capability qualified to perform a cluster of jobs of the same level. It is quite likely that in Japan employees with a particular qualification are engaged in jobs with different levels of duties and authority. The percentage of seniority-based compensation decreases from about 70 or 80 to 30 or 20 percent as employees move up from the lowest to the highest qualification rankings. Qualification-based compensation includes annual pay raise commensurate with the waiting period which implies that qualification-based compensation also is related to seniority to some extent.

Human Resource Management in the US-Sited Japanese Subsidiaries

US-sited Japanese subsidiaries must conform to American human resource management system to some extent and pay much concern for American values. If Japanese managers want to introduce Japanese human resource management system, they should modify their system to be accepted by American employees.

Hiring

When Japanese subsidiaries start their operations in the US, they need experienced employees who can do the jobs immediately, and so they recruit experienced workers in the local market, and recruit managers and engineers region or nation-wide. Japanese subsidiaries occasionally recruit new college graduates for managerial candidates, but it is not as yet the common practice.

Often states or counties sponsor training new workers and job applicants. Japanese subsidiaries take this advantage and recruit employees from newly trained workers or request states or counties to train new hires.

When job vacancies take place, most Japanese subsidiaries promote employees from within, using job posting system. When they have no qualified candidates from within the company, they recruit qualified employees from outside, and also they recruit entry-level workers, which is the common practice in American companies.

In selecting applicants Japanese subsidiaries consider job skills as the most important criterion. Most of the Japanese subsidiaries regard willingness to cooperate and a rounded personality as important essentials in selecting applicants. In the US where independence and aggressiveness are considered the most important traits of talented employees, much emphasis on cooperativeness and rounded personality often will result in the failure of attracting the top-group college graduates, especially MBA graduates from top business schools.

Japanese companies still are reluctant to recruit MBA holders which will result in the scare know-how to hire and utilize MBA holders in Japan and also in overseas operations. Most of the MBA graduates are talented and assertive. They are useful resources for the management of Japanese subsidiaries and for hybridization of American and Japanese management systems (Hayashi, 1985, pp. 145-160).

Another problem for Japanese subsidiaries to solve is equal employment issues. Japanese subsidiaries have the tendency to be reluctant to be involved in equal employment issues with long historical backgrounds in the US. One of the reasons is that Japan is a homogeneous society and so very few Japanese managers do not know how to respond to ethnic complexity in other countries. Another reason is that Japanese subsidiaries tend to focus on economic effectiveness and to be less concerned about other aspects of management, such as equal employment issues, public relations, and contributions to communities. Advancement of equal employment and contributions to communities are essentials in the US and other countries as well.

Orientation and Initial Training

The US-sited Japanese subsidiaries send core operators to the parent companies or factories for initial training during the set up phase. In Japan, American employees learn the production of products, production management, organizational culture and Japanese culture as well. After the initial training in Japan, core employees continue to learn on their jobs from Japanese engineers and technical instructors.

Japanese subsidiaries inculcate American employees organizational culture through morning meetings, picnic parties, Christmas parties, luncheons with managers and so forth. However, most of these means to better communication between managers and employees are used in some of the American companies and are not Japanese patents.

Job Rotation and Development of Multi-skills

Frequent horizontal job rotation across functional boundaries in order to develop generalists are not welcomed by American managers and staffs who would like to be specialized in a particular function. Job rotations across functional boundaries will lead to lower capabilities as specialists and will result in decreased chances to move to other companies for better compensation and jobs. In addition, when managers move to a new department without prior experiences, they should learn how to do jobs from their subordinates, which is not welcomed by both the managers and the subordinates. The managers transferred to different functions may lose their pride of being leaders and the subordinates may believe they can take the managerial positions instead of teaching jobs to the superior.

There are very few Japanese subsidiaries which practice systematic job rotations across the functional boundaries. Most of the American managers, engineers and researchers are reluctant to and often refuse the Japanese management's request to move to different jobs because they do not want to lose the advantage of being an expert in a particular function.

On the other hand, American semi and unskilled blue-collar operators are receptive to horizontal job rotations because they can acquire more job skills. Multi-skills imply better sales points for them and they can receive higher wages in the near future. Using job posting systems, Japanese subsidiaries inform blue-collar workers about job vacancies and encourage them to widen their skills through job rotations. In Japanese subsidiaries internal mobility is based on individual's desires to accept job rotation while in Japanese companies job rotations are conducted by the company's decisions, that is 'transfer order'. In the life-long employment system with few chances to move from firm to firm, Japanese employees cannot help accepting the company's transfer order. However, this practice cannot be applied to the US-sited Japanese subsidiaries where job hopping takes place quite frequently. On the other hand, if there is enough motivation, American blue-collar workers accept Japanese job rotation systems for the development of multi-skills.

Most Japanese subsidiaries have been trying to advance multi-skills among operators, and some of the Japanese subsidiaries established multi-skill pay systems which are designed to motivate American employees to acquire multi-skills to perform all jobs in the workplace where they are placed. Extra cents per skill are added to hourly basic rate each time when they have learned extra skills.

Many Japanese subsidiaries provide tuition aid plans to employees who want to improve job skills and develop their careers. The tuition aid plans of Japanese subsidiaries are quite similar with those of American companies.

Promotion from Within

In many American companies job posting systems are frequently used for promotion of employees though some companies still hold the privilege of at-will-employment for employees' promotion. One of the purposes of job posting systems is to regard employees' desires as being of paramount importance in personnel decisions, and another is strong concern for advancing equal employment for minorities, women, and other disadvantaged people, that is 'protected groups'.

Most of the US-sited Japanese subsidiaries follow the American practices. They use job posting systems for promotion of American employees. In other words, the individual's desires for promotion is the primary concern and then candidates are selected on the basis of capability and performance as well as concern over equal employment opportunities for protected groups. Promotion of Japanese expatriates is completely a different story. Promotion of Japanese expatriates is primarily decided by Japanese headquarters based on Japanese personnel appraisal and promotion practices.

In most of the Japanese subsidiaries, promotion across the functional boundaries is not as popular as in Japanese companies. Most American employees are reluctant to move to different functions, which implies that they will lose the merit of their expertise.

Succession/replacement programmes which are the popular practice for managerial promotion in large American companies is not popular as yet in Japanese subsidiaries since succession/replacement programmes are not popular in Japan and besides most Japanese subsidiaries in the US are employee-wise not as large as American companies.

Personnel Appraisal

In American companies personnel appraisal focuses on performance and capability, although the higher employees move up the organizational ladders the more emphasis is given to leadership, assertiveness, communication capability, interpersonal relations, and harmonious personality. However, these personality-related factors still focus on managerial capability to manage many subordinates to perform organisational objectives. Cooperativeness is considered as an important factor in personnel appraisal, but not such an important factor as it is in Japanese companies. Americans feel it is unfair for the factors not related directly to jobs to influence superiors' judgements for compensation and promotion of subordinates.

In Japanese subsidiaries, Japanese managers tend to regard attendance and attitudes of American employees as important factors in personnel appraisal according to their values and customs in Japan. Japanese managers tend to highly evaluate those American employees who do their jobs beyond what they are told to do by their superiors or job manuals, and volunteer to help peers' jobs. Japanese managers tend to highly evaluate those American employees who do not use up annual paid holidays. They tend to expect American employees to be loyal to the company and also Japanese team work which means employees help each other spontaneously.

In the US, Japanese managers' tendencies to emphasize long work and less holidays and also willingness to work beyond the assigned duties are not welcomed and even considered unfair. Therefore Japanese managers must evaluate employees primarily based on performance and capability in jobs.

Seniority

In Japanese subsidiaries Japanese managers tend to transfer and promote American employees according to their flexible seniority concept while American managers tend to use seniority as being the prerogative of long-tenured employees in the cases of lay-off, transfer, promotion and so forth. In Japanese companies seniority is used as the loose and flexible criterion to determine lay-off, transfer, and promotion. Japanese companies tend to lay-off older workers of higher wages rather than younger workers of low wages. They tend to transfer older workers to their subsidiaries to reduce work force redundancy. Seniority in Japanese companies is used advantageously for older workers in promotion and again is used disadvantageously against older workers to outplace or lay them off. As mentioned above Japanese companies select employees for promotion through the combined evaluation of seniority, ability, performance and attitude which gives managers more freedom and flexibility in promotion decisions.

In most Japanese subsidiaries wages are managed by job evaluation which is the common practice in the US. Exceptions are that some of the Japanese subsidiaries pay the same wages to office and factory workers with the same qualifications such as 'associate' or 'technician'. The concept underlying this wage system, is similar with that of qualification-based compensation in Japan.

Job Security

No Japanese subsidiaries guarantee job security formally by employment contracts or policy documents. Japanese managers are afraid that job security will result in the increase of low performers with less chances of being scouted, followed by low productivity. Japanese managers are careful not to use even the word, such as 'permanent employee' which often implies permanent employment.

However, most Japanese subsidiaries have been trying to avoid the lay-off of employees. Because many American workers believe in the myth of job security in Japanese subsidiaries, and lay-offs in Japanese subsidiaries will bring much more negative impacts than those in American companies on dissatisfaction and low morale of employees and will result in increasing turnover and low productivity. A Japanese manager even mentions that American workers feel it more difficult to work in Japanese subsidiaries than American companies because of the poor communications of Japanese managers, higher pressure to maintain product quality and vague job definition; and so in order to retain and motivate American workers, Japanese subsidiaries should avoid lay-offs as much as possible.

Concluding remarks

US-sited Japanese subsidiaries have many problems to solve in the cross-cultural interface in the increasing globalization and localization of corporate activities.

At first, long term development of human resources based on life long employment, which is the most critical feature of Japanese human resource management, has not been realized in most US-sited Japanese subsidiaries since very few Japanese subsidiaries keep the labour turnover of American employees as low as that in Japan. In order to improve the retention of American employees, Japanese subsidiaries ought to improve factors related to labour turnover such as compensation, work ways, duties and authority, communication, and participation in major decision making such as strategies and investment.

Secondly, Japanese subsidiaries ought to modify management and personnel policies which affect the activities of American managers. The collective and egalitarian features of Japanese human resource management such that all employees including managers wear uniforms, take meals together at the company cafeteria, join parties and trips are generally welcomed by American blue-collar workers and seem to be successful. However, American managers often resent this egalitarian feature of Japanese human resource management in such situations as egalitarian seniority-based personnel appraisal and slow promotion. American managers are puzzled by the vague definition of duties and authority, and resent frequent interventions of Japanese managers in their decision making, especially those Japanese who are ranked lower than them. Also they are dissatisfied with having no chances to voice their opinions on strategies decided solely by headquarters in Japan.

Thirdly, the contributions of Japanese subsidiaries to communities in the US are not enough to satisfy community people. Japanese subsidiaries ought to make more efforts to cooperate with the communities to solve difficult problems such as improvements in public education, equal employment, drug abuse and so forth. Japanese headquarters ought to give their subsidiaries more autonomy to allocate the resources to public relations and community issues such as time, money and personnel. Japanese headquarters should add to their training course for expatriates such courses in order to change the Japanese attitudes towards the community and motivate expatriates to volunteer for the improvement of community issues, and also should consider the expatriates' voluntary activities for the communities as one of the important activities of the subsidiaries.

Finally, in order to cope with globalization and localization, Japanese companies ought to modify their tendencies to consider as most important the interests of Japan and their companies and the focus on economic effectiveness, which brought Japan economic growth and prosperity in the past decades, and ought instead to innovate more human centred philosophies and management systems.

References

Berkstresser, G A & Takeuchi, K (1983) Collectivism in Japanese Industry: A more Realistic Perspective. In *Journal of Humanities and Natural Sciences* No 64, pp 1-14, Tokyo Keizai University

Hayashi, K (1985) *Cross-cultural Interface Management: Japanese Management Overseas* (Japanese ed.) Yuhikaku Pub.

Industrial Cultures in the Age of the Global Economy

Yuji Masuda

Industrial Culture in Global Economy

The modern industrial culture is at a turning point and also being challenged. As economic activities are internationalized and companies try to find a larger market abroad and participate in local economic activities in each country, friction between different industrial cultures is likely to occur. In the age of interdependence caused by internationalization in various fields such as politics, economy, culture, and defense, contact between different cultures in many ways increases in causes of friction.

Contact between different cultures generates misunderstanding and distrust because of difference in languages, religions, ideologies, customs, etc., and these differences give rise to discord. Each culture has its own originality, and claims justice. Of course, each culture has identity (originality) and no culture is superior or inferior to another culture. Cultures do not have differences in levels but in characteristics. However, characteristic differences may result in disputes. Actually, cultural differences generate international disputes all over the world.

The cultural friction in question is not always a dispute between traditional cultures. As seen in the US-Japan discussions on structural barriers, the current friction has been caused by the modern 'industrial culture'. The industrial culture is a keyword at present.

The industrial culture herein referred to is the sum of the values, labour customs, production systems, etc. that have been formed by the geographical conditions and historical background of an economic society and it indicates the environmental conditions of the entire social economy surrounding them. Each country has an individual industrial culture and modern industries have been established on the industrial culture.

Japan is often regarded as a special country. Presently, Western people claim that Japan is so as a result of the increase in Japan's trade surplus and the US-Japan trade imbalance. Japan may be viewed as special because it is different from other countries. In contrast, Western countries are special countries having special industrial cultures. The claim that Japan is special is based on Western standards.

Surely, Japan has strange social customs. As Japanese companies find a larger market abroad, the Japanese style of business is of great concern in Western countries.

The industrial culture consists of sectors such as social values and labour morals as the first sector, technological development and creativity as the second sector, and the so-called culture that is user/consumer activities as the third sector.

The first category is associated with evaluation on man's existence in the society (i.e. From what does values derive? and Why do people work?). Max Weber claimed the 'protestant's spirit' for the reason that capitalism was formed in Europe and became the world-wide system. Also, it has been discussed that Confucianism has supported the economic growth in East Asia in recently years. Somehow, these matters indicate that the labour morals define the industrial culture.

The second category is associated with the working system, which is the permanent employment and seniority system in the Japanese style business management. These systems are featured in the Japanese type industrial culture. It should be noted that they have positively worked in business administration and smoothly promoted a succession of techniques and skills in industries. However, whether they have universal characteristics is suspect. Further, this sector is associated with the creation of a new social value and technology. It is often said that the 'Japanese lack creativity'. Indeed, the Japanese have not achieved positive creative technological development. This may be due to the world-historical situation of Japan.

The third category is the total living culture which is the lifestyle, consumer activities, etc. formed in the social economic relations. It corresponds to the production culture and can be a strong base for the social economic structure.

Industrial Cultural Determinants in the American Machine Tool Industry

In an empirical study on the industrial culture, the process of reducing the competitive force of the American machine tool industry indicates a typical example of dynamism of the industrial culture.

Generally, it is considered that international competitive force is associated with factors such as technological power, exchange rate, etc. In addition, the status of the industrial culture seems to affect international competitive force greatly. For example, let us consider the machine tool industry, which supports the international competitive force of industries. It is public knowledge that the machine tool industry in the United States has rapidly lost its power. There are almost no traditional machine tool manufacturers any more.

It may not be adequate to attribute the cause of recession to the exchange pressure. As internal factor, the *cultural* factors configuring this industry should be considered.

In this case, *culture* is a very wide concept. The organization system of a company can be considered to be a culture. Also, the thinking method and labour organization system are cultures. In other words, culture indicates not only traditional cultures but also the elements that configures the modern industrial society. Therefore, the reduction of the competitive force in the American machine tool industry can be explained by taking the cultural factors into account this way.

It can be said that the status of the 'culture' defines the industry. The industrial cultural determinants do not refer to the determinism but the mutual relation (i.e. mutual 'effects") between industry and culture.

The reason for the recession in the American machine tool industry is that the relationship between the industry and culture became unreasonable. This industry is the base for industrial activities. It still plays an important role in activation of the industry and culture and it is a determining factor. How this industry is re-energized for the benefit of industrial society is an important subject. It means that it will be necessary to restructure the relationship between the industry and culture, and political proposals for this purpose are needed.

In restructuring the relationship between the industry and culture, the viewpoint taken to associate them is a political subject. If the technology-oriented viewpoint takes the precedence over the viewpoint of living and working, restructuring becomes technocentric. For re-activation of a powerful industrial culture, it is required to put human labour and living at the centre in structuring the social system, production system, and other systems.

Restructuring of the Industrial Culture in EC Countries

Restructuring of the industrial culture is taking place internationally. EC is deliberately working towards unification in 1992. FAST announced ' Competitive Force of Europe in the 21st Century' at the end of 1989. The subtitle is 'Unification of labour, culture, and technology' which makes proposals about systems of human work system and the direction of research and development in the advanced industrial society.

The report indicates the disadvantages of using production technology and inventions because Europe lags behind the United States and Japan in the high-tech fields. However, Europe will not remain so; EC has the capability and tradition of developing scientific technologies and products matching its cultural, geographical, environmental, and economic status as the largest trade block in the Western world.

From the 1960s onward, EC countries have been challenged as a super power by the United States and Japan. However, they have a long cultural tradition and recognized priority over the United States and Japan as newly industrialized countries. As a counter-challenge, EC must be unified. Business cultures in the United States and Japan have concentrated their efforts on integration and expansion, but the business culture in Europe focuses on collaboration and versatility. Any culture has advantages and disadvantages.

Cultural wealth, versatility, and creative spirit in Europe have made it difficult to develop forcible and concentrated approaches to industrial development. However, these disadvantages are becoming advantages. Of course, it is difficult to measure the economic merits of different cultures based on different status and different time. However, in the broadest meaning, the collaboration, advanced cultures, and creativity bring superiority. European countries surely have advanced traditional cultures and arts, which have been diffused throughout the world and have often energized the modern industrial culture. EC countries are trying to make this cultural transition reactivate the industries and restore the competitive force

Enhancement of Competitive Force by the 'Anthropocentric' System

Europe is behind the United States and Japan in the high-tech fields and is trying to develop a EC's strategy within the EC, into the 21st century to catch up with these countries. As a practical measure, EC should positively promote cooperative research and development project in high-tech fields such as information technology, bio-science, new materials, etc. EC countries are cooperatively trying to compete with the United States and Japan. As a result, it is expected that technological development capability will be enhanced and the technological level advanced. But cooperation is merely a necessary condition for reinforcement of the competitive force.

To acquire Europe competitive force, the sufficient condition is to promote restructuring of European industries. For this purpose, the social infrastructure, new product development, educational reorganization, and changes of the production system are essential. Since the United States is too technocentric and Japan is too corporate-centered, EC should emphasise the 'anthropocentric' labour and technology development. Anthropocentric technology tries to utilize high technologies based on man's skill and creativity. Through anthropocentric technology, real symbiosis of labour and technology is possible if the technology is of the adequate form.

EC has created the following targets for the technology development policies and research subjects:

The first is creation of Europe's unique functional system of scientific technology corresponding to the world economy, European single market, and variations of population organization. The second is structuring of the production system incorporating characteristics of flexibility and flat hierarchical structure. The third is enhancement of the competitive force by product quality, versatility, etc.

For this purpose, optimum matching between the technology and the corporate organization and between the product and production organization is required in order to develop a system that will accommodate the society-technology system. Also, it is necessary to analyze local differences in the production technology, products, etc. in the EC area, clarify the society-culture factors, and to use them in system development. EC is trying to restore the vitality and competitive force through the creation of system development methods utilizing the social characteristics.

In the technocentric relationship between man and technology, technology plays the active role, while the human is passive, and corresponds to components of systems. Since these machine systems have characteristics such as prediction, repetition, and quantitativeness, they often seem to be scientifically valid. In contrast, they eliminate intuition, subjective judgement, tacit knowledge, imagination, and will.

In the anthropocentric technology system, technology itself supports man in order to let the human utilize these capabilities. Technology now supports the human's technological capability and creativity. The concept of complete automation will never be established in the context of EC's technological development, and it will be eliminated. Instead the creation of advanced production systems structured on an anthropocentric basis is the method for EC's technological development. As a result, EC can acquire international competitive force.

Japanese Companies Finding Their Way Overseas and Global Industrial Culture

Both the industrial culture and economic globalization will further direct attention. Japanese style business customs cause local friction as a result of finding a larger market abroad by Japanese companies. In more than half of companies operating in advanced countries and almost half of companies operating in developing countries result in friction because of differences in business custom. It is not merely the problem of the corporate culture in a company. Important matters are often determined in the party after working hours in the Japanese style business. It is really a strange system from the international standpoint and the authority in decision making seems ambiguous. Formal and informal matters are mixed and decisions on a formal matter are often made in an informal place. Also, it has been said that Japanese manoeuvre 'true intention' and 'principle' skillfully. Therefore, foreigners can hardly understand the situation.

Most Japanese companies have these characteristics, which stand out when the interface with the local industrial culture is established as a result of finding the way abroad. At the same time, Japanese companies appear to be closed. The ambiguous decision-making process and administration that tends to be dominated by the headquarters enhances the recognition that the Japanese corporate culture and industrial culture are strange. Thus, the cause of friction is attributed to the 'Japanese style business management'.

Maybe, members who belong to the same industrial culture work together and can share information easily in Japanese companies. Information can be implicitly transmitted in the mind-to-mind communication without explicit comments. In Japanese society, implicit expressions are often used for information transmission. Japanese society and corporate structure are very efficient social systems for information distribution.

In contrast, Japan's social characteristics which have a completely different aspect in the international scene are often regarded as a peculiar system inevitably. Friction and problems caused overseas by Japanese companies is a reflection of Japanese style information sharing and characteristics. Friction maybe also caused because different cultures contact each other. Experience of Japanese companies overseas has an aspect different from the experience of Western companies.

The present world economy system is dominated by the Western industrial culture and corporate culture. On the other hand, Japanese style business management is brand-new in the international scene. However, it is true that the Japanese style business management has had an international competitive force since the latter half of the 1970s. On of the reasons is efficiency in information distribution among members with same quality, but efficiency is not guarantied among members with different quality. In the international scene, this fact causes a big problem.

The modern industrial culture based on people's nations is in the transition period. Along with economy globalization and economic unification advance, a global industrial culture will be formed.

Destruction of Past Barriers and Direction of the Industrial Culture

The corporate culture and industrial culture are widely changing toward the 21st century since 1990s. Presently, they are in the transition period. So far, corporate activities have been allocated based on the single culture zone. From now on, corporate activities will take place in the globalism exceeding the barriers of the single culture.

The world economy system in 1990s is in the transition period toward the 21st century while destroying various barriers in the past and creating a new structure. In this case, the world economy system eliminates the barriers of two economy systems; the traditional capitalist economy and socialist economy. So far, the socialist system of economy has been structured on a basis different from capitalism. However, the socialist system based on the economy plan led by the communist party is being destroyed.

In 1989, the situation of Eastern Europe that had been under the dictatorship of the communist party was drastically changed by the economic reorganization led by USSR's *perestroika*. This reorganization brought not only economic reorganization but also political and social reorganizations as a multi-aspect reorganization process and it destroyed various barriers.

First, the wall in Berlin was destroyed and the physical barrier between West Germany and East Germany disappeared. The 'Wall' as a symbol of separating the East and west from each other became a past thing. Consequently, the economic barrier is being weakened and the East and west are coming closer to each other in economy. Of course, weakening of the economic barrier is occuring not only between the East and West but also in the Western world rapidly as can be seen in the plan of EC Unification. These facts indicate that new economic unification is going to be promoted by eliminating the wall between economic systems and the wall of people's economy in each country.

Further, promoted by disappearance of the economic barrier, the technological barrier is going to be weakened. So far, advanced countries cannot export high technologies to the East because of COCOM (Coordinating Committee for Multilateral Strategic Export Controls). It was part of the 'containment' strategy of the Western world lead by the United States against USSR, which began in the age of the cold war. Now that the United States and USSR are coming closer to each other the United States is going to remove the COCOM wall made by itself.

Ironically, the last barrier is the races. USSR consists of various races and each race has formed the people's republic.

As a cultural problem, the racial problem exists in the modern capitalist system that was structured on peoples. This barrier also represents the problems in the modern world in a concentrated way.

Presently, the physical, economic, and technological walls that have separated the two economic systems from each other are going to be removed. How will the new industrial culture be structured?

Two Social Economic Systems and the Industrial Culture

Presently, we are in the age when barriers are being destroyed and a new order formed. However, it is anticipated that its process will contain lots of troubles. Economic and technological barriers as systems made by man can be overcome through reorganization of systems. Particularly, the socialistic system is greatly changing.

USSR's perestroika has had an impact on Eastern Europe. The dictatorship of the communist party that had dominated the people who were requesting democratization and liberalization was destroyed. Such movements were concentrated in a short period in 1989. In contrast, USSR which would not abandon the dictatorship of the single party seemed to be behind the trend of democratization and liberization. However, Gorbachev made a historical decision at last.

The rapid current of political reorganization in Eastern Europe flowed back to USSR. Consequently, the authority and dominating power of the USSR's communist party decreased drastically. For improvement, the only way was to abandon the dominating power and to employ the competition system at the political level through introduction of the multi-party system instead of reinforcing the dominating power in accordance with Clause 6 of the Constitution. President Gorbachev made this decision.

This system reorganization does not merely reorganize the political system based on Stalinism. Further, reorganization of the economic system will be brought about. It means various types of social ownership, unification between the individual's benefit and social benefit, and wide utilization of the market functions. This fact is the economic basis for 'humanistic socialistic democracy' and indicates the trend of employing past results from the Western socialistic democracy.

What was the socialistic economy? Where is it going? These are in question. At the same time, capitalism is in question. Maybe, the two economic systems will stand on the same basis in terms of utilizing the market mechanism. In this case, treatment of the industrial culture which ensures versatility is important.

Presently, the market economy of capitalism is in the age of high technologies internationally. Under the scheduled economy, goods and services are produced according to commands. In this stage, not the quantity of production of iron, etc. but increasing the added value is important. However, in the socialistic system at present, it is hard to measure the added value. Further, as the industrial society places its weight on knowledge production such as research and development, design, etc., measurement and scheduling become more difficult.

Although the industrial culture uses a labour moral as an important component, socialism failed in creating its original moral and the scheduling theory in the age of knowledge production has not been established. It is not merely solved through introduction of the market theory. Presently, it is required to structure the original industrial culture for economic reorganization.

Political Economic System in the 20th Century and New Formation of the Industrial Culture

Presently, the political economic system in the 20th century is in the stage of review. Particularly, its basic frame is in question. The basic frame herein referred to is the theories of democracy, racism, and socialism established after World War 1 and they are required to change. So far, the political economic system in the 20th century has employed the three-layer structure consisting of democracy, racism, and the sovereign nation. The modern society has created a modern nation in the capitalistic market economy based on each race. However, the cold war after the World War II broken independence of each structure.

In other words, the way of 'partitioning' is changing. After the World War II, particularly during the cold war, the United States and USSR gave a significance to each part of the world and conducted the world. However, the Wall in Berlin is being dismantled physically and mentally and partitioning is being changed. The barrier of technological exchange is being removed, economic exchange is activated through introduction of the market economy system into socialism, and the economic barrier is going to be dismantled. The economic unification program is progressing in East and West Germany.

Associated with the change of partitioning, the meaning of confrontation between the East and West is changing significantly. These words have meant the confrontation between ideologies. However, it may mean the cultural difference between the Western world and Asia as the original meaning of the East and west. Market economy and democracy are going to extend beyond the Ural Mountains and Eastern Europe is entering the 'West' as Middle Europe. On the other hand, China, Japan, and NIES in Asia are being reorganized in the 'East'.

If partitioning between the East and west has the geological meaning, recognition of the Western world versus Eastern world may occur. Then, the 'strangeness' of the Eastern culture, particularly the Japanese industrial culture, will be insisted upon.

Formation of such recognition is unfortunate to both Eastern and Western worlds. If recognition is fixed, 'East is East and West is West'. Conversation between the East and West becomes impossible and mutual confrontation and distrust are easily created.

Concepts of the race, democracy, and sovereign nation are shaking. The frame that created the 20th century is going to be rebuild. What is important is not to regard races in the fixed manner but to respect each racial culture and enable the exchange. Under such circumstances, the political economy system to be structure in the future is unknown. However, things are getting serious. The world economy system must be designed and restructured as soon as possible.

Since internationalization of people's economy has advanced, economic interdependence has increased, and economically isolated socialistic countries are taking the open policy, alteration of sovereign nations has begun. Along with the flow of materials and money, people are moving internationally. The political economy system in the 20th century is fading away. The partition for each sovereign nation is becoming vague and ambiguous. As barriers are destroyed, people have faced the racial walls that had formed modern nations. Racial affairs in USSR indicate one of the problems in a modern nation. Such racial affairs are occuring throughout the world.

The ending of the East-West cold war is not indifferent to Japan. In this stage, selection by Japan and the action to be taken are being asked internationally. At the beginning of

1990s, destruction of the after-war world frame and cold war and the change in the relationship between the East and west will define the structuring of the world political economy system in 1990s and the 21st century. Also, selection and embodiment of the political economy system by Japan are asked.

The common matters under these conditions is that people, materials, and money are moving beyond the frame of each people's nation and corporate activities are multi-nationalized and connect various economic zones. People having different values and labour morals take economic actions in various organizations. In the modern industrial society, values and morals that had been fixed in certain frames cross over. Under such circumstances, the corporate organizations are required to accommodate the different values and labour morals.

Each industrial culture has identity and a long history. However, at the same time, the identity must not be fixed and exclusive. It must accommodate the age while contacting other industrial cultures and creating a new form. Consequently, exchange between different industrial cultures is allowed. In the global age from 1990s to the 21st century, the international social economy system is going to be structured based on various industrial cultures complying with the market economy, racism, and races. The industrial culture has entered new horizons.

Section III:

Human-Centred Shaping of New Technologies: The Praxis

Theoretical Reflections from Experiences with Computer Aided Design and Manufacturing

Richard J Badham

Introduction

As was apparent in the 1990 IIASA 'CIM: Revolution in Progress' conference in Vienna, the approach to computer integrated manufacturing (CIM) taken by 'human centred', 'skill based' or 'anthropocentric' researchers is still highly controversial.[1] Moreover, within the growing number of human centred design researchers, the exact nature of 'anthropocentric' or 'human centred' technologies is still a subject of concern and dispute. This was clearly expressed in recent discussions in the first and second Anthropocentric Technology and Society Newsletters.[2] What is less controversial, however, is the increasing awareness that:

(i) there are different *forms* of computer integrated manufacturing technologies,
(ii) these forms are *shaped* by socio-political as well as technical considerations, and,
(iii) there exist significant *choices* to be made between more or less desirable forms of computer integrated manufacturing.

Where fundamental disagreements occur is over the *extent* to which the different forms of CIM can be classified into 'technocentric' versus 'anthropocentric' or 'human centred', the *degree* to which the choice between these systems can be adequately explained as resulting from the fundamental conflict between 'Tayloristic' and 'post-Tayloristic' technological paradigms, and *how far* the choice between desirable and undesirable forms can be captured by the technocentric/human centred dichotomy.

The purpose of this chapter is to examine these general theoretical issues in a discussion of the design criteria, design processes and stratetegies put forward by human centred manufacturing system designers. In order to avoid an overly abstract analysis, the discussion is undertaken through a consideration of organisational experiences with computer aided design and manufacturing technologies (CAD/CAM), and the general issues these raise for appropriate human centred design criteria, processes and strategies. The examination of CAD/CAM is based on an international review of CAD/CAM experiences and an investigation of 10 CAD/CAM user organisations in Australia.[3]

CAD/CAM and the Factory of the Future

One of the significant features of CAD/CAM is that it has provided the cornerstone of much of the contemporary imagery of the 'unmanned factory' that is directly challenged by the human centred approach. This has been explicitly observed by Professor Rosenbrock in his recent discussion of the longer term significance of his project to design the human centred lathe.[4] The images of the automatic factory are, of course, far from new. In ancient Greece the philosopher Aristotle remarked that, 'If every tool when summoned, or even of its own accord, could do the work that befits it, just as the creations of Daedalus moved of themselves, or the tripods of Hephaestos went of their own accord to their sacred work, if the weavers' shuttles were to weave of themselves, then there would be no need either of apprentices for the master workers, or of slaves for the lords'[5]. Continuing this theme, Antipatros, a Greek poet of the time of Cicero, and an apparently budding classical feminist, hailed the invention of the water-wheel for grinding corn, an invention that is the elementary form of all machinery, as the giver of freedom to female slaves, and the bringer back of the golden age.[6]

In slightly more recent observations on CAD/CAM such as those of the Economist magazine in May 1987, we are asked to 'Imagine an engineer sitting at a computer terminal punching in data for the design of a new product and sketching freely with a lightpen on the screen before him. Happy with the design, he presses a button and the details are passed electronically to another computer running software that checks to see whether the design's stresses and strains are within prescribed limits. The information then zips along to a third computer which generates instructions that command the tools in the workshop to machine, assemble and store the engineer's product ready for distribution - all done automatically, without hassle, delay or hefty handling, and all before the morning's coffee break. One more satisfied customer. Welcome to the 'factory of the future'. The enthusiasm of such images has led to comments such as the following from one Unimation company president, 'I don't think a guy will be able to go to his country club if he doesn't have a CAD/CAM system in his factory. He's got to be able to talk about his CAD/CAM system as he tees off on the third tee - or he will be embarrassed.'

Such simplistic imagery of CAD/CAM use is also applied, however, in the contrasting views of a number of Marxist labour process writers and socio-technical theorists of the integrated factory. According to the first image, CAD/CAM is used primarily in order to separate conception from execution in the workplace, de-skill traditional occupations, and centralise functions and control - consequently, shopfloor craftworkers and indeed the manufacturing sector of the firm loses functions to design, and within design the traditional skills and functions of draughtspeople are lost to the CAD/CAM system and the engineers running that system. According to the second image, of the socio-technical integrated factory, environmental turbulence, technological uncertainty and interdependence all lead to the general raising of skills of system operators, increasing responsibility of both design and manufacturing workers, a growth in team collaboration, and a reduction in traditional hierarchies.[7]

However, in the international review of CAD/CAM, it was discovered that no simple statements could be made about a general centralisation of functions and occupational 'deskilling' with CAD/CAM nor about a general trend towards the establishment of a socio-technical model of an integrated factory characterised by greater skills, increased teamwork, and reduced hierarchy. The reason for the inadequacy of these different models was primarily twofold: the variability of CAD/CAM configurations within user firms; and the complex organisational conditions that lead to the selective development, use and impact of these configurations. These two factors merit further consideration as they raise key issues for both the definition of human centred technology and strategies for designing and implementing such systems.

The Variability of CAD/CAM Configurations

By discussing 'the' consequence of introducing 'CAD/CAM', previous studies often failed to distinguish between the impact of not only levels of sophistication in the basic hardware and software, but also of such characteristics as different configurations of the system (and associated allocation of functions to control levels, people or systems), forms of system development and customisation, and alternative human/machine interfaces.[8] Thus, for example, changes or updates in CAD drafting or NC programming software had a direct effect on the degree of frustration caused by system delays or the ability of machinists to use the programming tools. In addition the organisational locations of system functions and capacities and the nature and effectiveness of the communicative links between system elements.had a significant effect on inter-departmental and inter-firm activities. This varied from the effects of decisions on whether or not to allocate any CAD/CAM workstations for use by manufacturing personnel, to the data entry capabilities and level of programming language used by operators on NC machines, and the success of firms in shifting functions between firms by creating effective IGES links.

Once this variability is recognised, it is apparent that the purchase of a particular system involves the acquisition by the firm of a *'technological resource'*, it does not become a *'working technology'* until four other actions are initiated. Firstly, operators need to be found and skills need to be learnt to operate and maintain the system. Secondly, the technology has to be located and integrated both technically and operationally with the rest of the firm's technological or production system. Thirdly, the full exploitation of the standard technical capabilities of the system require the establishment of data bases and effective data entry, and the customisation of user-interfaces. Fourthly, the development of the system requires updating of data-bases, the writing of new programmes, interfacing and integrating in new ways, improvement of user interfaces, discovery of new uses for the system or more effective ways in which it can be used etc. Only the potentialities of the system that are exploited and developed have an 'impact' on the organisation.

For the effective use of the system in design and draughting, for example, the development of macros and menus and standard parts libraries is crucial. Yet in many firms this has not occurred, and the consequent 'impact' on jobs, for example the potential routinisation of drafting work, has not proved possible. Moreover, the extension of CAD/CAM use into many production planning and manufacturing areas requires the creation of 'families of parts' or 'group technology' and their combination with other attributed data, but these features have often not been implemented. Similarly, for effective CAD/CAM systems the development of suitable post-processing software is essential to ensure effective communication with NC machines, yet many of these links, providing the possibility of shifting functions 'up the line', had not been developed. Also in the communications area, the greater the sophistication of communication links between firms the more functions can be shifted 'down the line' either from the design area of the customer firm to the manufacturing sub-contractor or vice versa. In many cases this has not been achieved, with firms recognising the inefficient duplication of data entry at either end but the technical preconditions for organisational change often not being achieved.

This variable development of CAD/CAM features should not, however, be simply understood in terms of the *level* of development of systems. Technologies are selectively customised and developed depending upon the organisational interests and production requirements of the firm. Systems cannot be adequately conceptualised in terms of integrated wholes that have reached a certain stage of development but are more appropriately seen as complex configurations, the elements of which are selectively used or not used, developed or not developed, depending on specific circumstances. In an important sense system components may be seen as relatively autonomous 'nodes' of the configuration, and their initial form, future development and use will be influenced by the organisational context in which they are embedded. These organisational contexts may represent, in a sense, cross-cutting rationalities that may be more or less in line with the original intentions of the designer or the 'effective' integration or working of the overall system. This is what makes such systems, in the words of Brian Wynne, 'unruly technology'.[9] At one level this has been manifested in a variety of particular technical conditions in CAD/CAM systems, for example: in some cases key interfaces have not been developed either within the company or between the company and suppliers/customers, this prevented certain reallocations of functions between design and manufacturing and between the company and sub-contractors/customers; in many instances CAD databases had been selectively developed, and software only developed and customised to a limited extent, partially influenced by and influential for established organisational structures in the firm, sometimes CAD/CAM workstations were located in a specialised section within design, sometimes distributed throughout the design section, and at other times distributed between design and manufacturing, with consequent implications for the nature of system use etc.

At another level, more general and systematic influences can be seen to influence the development and impact of the system. It is apparent, for example, that firms have different 'design profiles', i.e. the relative weight or priority they give to concept design, engineering analysis or detailed design within design, and, the overall priority given to design for production and design within production .The result of these different 'design profiles' is a greater attention by firms to particular areas of their production process, with consequent implications for the level of technological development, and skills, status and power of different sections within the firm. One United Nations industrial development study into computer aided design noted, for example, the different speeds of diffusion and impact of CAD within firms which were more design or detailing intensive.[10] More specifically, in the Australian study it was observed that Australian manufacturing is generally characterised by small size, low level of research and development, and a weak performance on other areas of design and innovation. The case study companies using CAD/CAM were, therefore, typically characterised by a low level of in-house design, and consequent lack of development of CAD/CAM in this area. This was most clearly apparent in the failure to adopt, develop and use computer aided engineering analysis packages, even in firms carrying out quite complex manufacturing operations or concept design work. This has direct implications for the design engineer/draughtsperson interface. As engineering analysis results in the creation of a product design data base that can also be used for draughting purposes, there are consequent pressures to increase either job integration or organisational integration. In the absence of the use and development of this capacity in the case study firms, there was a general awareness of the possibility of engineers using engineering analysis, and subsequently taking over some draughting work, but this was not perceived as a major issue.

Shaping and Use of Technological Configurations by Organisational Structures

Any discussion of the origin of such design profiles and their effectiveness in influencing technological development brings us to our second question, the organisational context that influences the shape and impact of technological configurations. With regard to the development of engineering analysis capabilities, for example, the low level of in house design in the Australian CAD/CAM user firms clearly affected the lack of significant advance in this area. In addition, however, some firms mentioned the influence on this process of skills and training in Australia. Firstly, there had in the past been little training in the use of computer aided engineering, and in addition engineers had been brought up in a culture that emphasised that 'mere' draughting was below the level of attention of engineers. Secondly, in some companies there had been explicit opposition by the draughting and technicians trade union (ADSTE) to any use by the engineers of the CAD/CAM system. These conditions combined to reduce the attempt by firms to develop and integrate the engineering and draughting capabilities of CAD/CAM systems, and there was consequently little pressure for the takeover of drafting functions by engineers.

Design profiles, involving a firm's prioritisation of design areas, may thus be influenced by a variety of factors including the skills available, ability to compete in particular market areas, industrial relations conditions etc. Such profiles are, however, clearly affected strongly by the nature of the product and the required production processes. Where CAD/CAM systems are used for electrical rather than mechanical applications, for example, the relative ease of detailed draughting and simplicity of the manufacturing process enhances the integration between design and manufacture, and there is consequently a greater use of system 'design rules' for automating this operation, and the performance of such integrated design/manufacture functions, by design engineers.

Despite the extensive literature on the rise of 'flexible specialisation', and the establishment of new 'production paradigms' or 'production concepts', the Australian study was unable to make a direct link between the use of CAD/CAM and companies located in cost-competitive/large batch/standardised markets and others in quality-competitive/small batch/customised markets. Many firms, for example, had product markets that cut across these distinctions, i.e. they were attempting to enhance component standardisation while increasing the rate of product innovation. Other firms produced large batches of products but the relatively great costs of production rather than design meant that they could afford greater experimentation in design and were not so concerned with direct economic payback from the system through enhanced draughting productivity. Moreover, CAD/CAM was used within a number of firms to produce a variety of different products, including tools and fixtures for use within the firm. In addition, the use of the system in a more or less routinised manner depended on the degree of innovation and change within the production processes as well as the product i.e. the degree of emphasis upon the updating or renewing of design software, production equipment, or organisational techniques.

As detailed in recent work by Saorge and Streeck in Germany, and Child in the UK, a strong influence is exerted upon firm's organisation and use of such technologies as CAD/CAM by the degree of process variability and product innovation - as indicated by batch sizes, length of production runs, price or quality based competition, and product range or variability.[11] Such influences exist, however, within the context of other important intra-firm and extra-firm conditions. At the intra-firm level socio-technical configurations are directly influenced by what Child has called the 'inherited characteristics' of a firm i.e. the size of the companies and plants, the stock of competencies and skills, traditional modes of operation and labour control, and the internal power structure.

Within the constraints imposed by their environment, organisations possess a range of 'strategic choice' and 'design space'.[12] In particular, as emphasised by McLoughlin and Clark, there can be great variation in the 'tightness of coupling' between senior management objectives and lower level management influences on system structure through the decisions they made on the implementation strategy and operation of new technology. It has been observed, for example, that there are substantial differences between more participative (and 'bottom up') or less participative (and 'top down') approaches that influence which objectives of which groups become embodied in the final working system. Senior management, for example, often has strategic objectives (such as improving product quality), middle line and financial managers frequently have operating objectives (such as improvements to product flexibility and reductions in plant running costs), and middle to junior line managers often have control objectives (such as amount and speed of performance information, control over workflow, and labour control)[13].

Similarly, in West Germany, the experience of the production control section of the Fraunhofer Institute for Production Systems and Design Technology has been that top management is concerned with overall efficiency and profit, middle management with cost reduction, and operational management with production control and worker motivation, planning department with the specification of planning details but also less repetitive and more creative work for planners, support departments with independence and higher status, foremen with independence from management/planning and control over workers and support departments, and workers representatives with labour force contentment, balancing blue and white collar conflicts, and some humanisation of work ideals.[14]

Continuing this theme, Eckart Hildebrandt, researcher at the Science Centre in Berlin and member of the CAPIRN network, has emphasised the central role of the '*social constitution*' of firms in determining the design and introduction of production technologies.[15] Defining social constitutions as the 'overall ensemble of the most important norms and rules permitted or effective in the plant that influence the employee's attitude toward work and behaviour at work',[16] Hildebrandt emphasises the central impact on production planning systems in German machine building firms of: plant size and status; relationship between planning and execution; the independence of plant areas; means of personal coordination; nature of parallel structures and contradictory principles; the position of the works council in the structure of plant pressure groups; and local culture and its influence on the integration of the factory.

In a comparative study by Sorge et al. of the UK and German machine tool industry, focusing on the use of CNC machine tools, a similar discovery was made, that '*socio-technical traditions*' influence the development and use of technology. The factors discovered to have a key influence were company and plant size, batch size, type of cutting and machinery, national institutions and habits of technical work, management and training; and socio-economic conditions relating to natural resources, size of product markets and rate of growth.[17]

The particular aim of the latter study was to go beyond an examination of intra-firm conditions to compare alternative national industrial cultures. In this regard particular emphasis was placed on the different qualification structures of the UK and West Germany, and organisation divisions between planning and shop floor, as a major influence on the location of programming with the new machines. In the work of Sorge and Streeck, this was extended further into an examination of the central role of 'national patterns of skill formation' in influencing both industrial relations characteristics and the product markets into which firms enter. In contemporary ongoing research on 'national systems of innovation' and 'national conditions of enterprise calculation' this is extended beyond the skill formation arena to include financial institutions, government activity and policy, market structure, capital goods user/producer relations, dominant national economic blocks and the composition of demand, as well as the industrial relations system.[18]

Human Centred Design: Criteria, Processes and Strategies

Many of the features of CAD/CAM development and use outlined above will be familiar to human centred CIM researchers. In particular, the interest of the human centred CIM movement in shaping technological configurations explicitly recognises the malleable nature of systems, and the incorporation of political and social elements in their structure. Moreover, in the work of writers such as Peter Brödner and Chris Clegg, there is a strong emphasis on relating systems design to firm's strategic considerations.[19] Particular stress has been placed on the use of Sorge and Streeck's classification of emergent forms of 'diversified quality production' and the importance of improving the integration between business strategy and system development. In addition, with regard to the integrative character of CAD/CAM systems, Peter Brödner and more recently the members of the ESPRIT 1217 group have directed attention to the idea of establishing *design islands*, and improving the integration between design and production using technologies such as the 'electronic sketchpad'[20].

In addressing these issues human centred design projects have concentrated on three areas:

(i) the establishment of practical and utilisable design criteria *of human centred systems;*

(ii) the outline of procedural guidelines for an effective human centred design process*; and*

(iii) a strategy for promoting human centred systems that involves the creation of working technologies or systems to be used as examplars *of human centred design to be used in shifting production* paradigms.

I would like to argue, however, that these three areas require reconsideration and development in order to take into account the complex and variable nature of technological configurations and the organisational processes responsible for the selective development of these configurations.

Design Criteria

In the area of design criteria, three key issues have been raised by the CAD/CAM discussion. Firstly, central attention should be paid in definitions of human centred systems to the 'processual' nature of technological configurations, and the establishment of a human centred process of system development. By this I mean more than the usual discussions of appropriate processes for the design of human centred systems, i.e. involvement of end users, use of rapid prototypes, employment of concrete scenarios etc. What is meant is the creation of a technical and organisational structure that facilitates further system learning and development, by facilitating and encouraging the continued collaboration and commitment of the system participants. As illustrated in various studies of technological systems as well as job design experiences, system continuation and development requires continually overcoming not only external opposition but the tendencies towards fragmentation brought about by the relative autonomy and different interests of system elements or participants. In the case of systems such as CAD/CAM, involving the integration of organisational sectors of the firm with a long history of independent technical and organisational development, this requirement becomes crucially significant. This is most apparent in the crucial significance of system development through the time consuming and complex establishment and updating of data bases, programmes and system interfaces

A second area of further research is due to the fact that, depending upon the strategic goals, organisational structures and information flows of different firms, the requirements of systems and consequently the form taken by human centred systems will differ. In particular, as observed above, firms with different 'design profiles' have different requirements for, and interest in, new design and manufacture systems. Moreover, firms employing less skilled personnel in more 'peripheral' industries have a different set of resources available to the human centred systems designer, and consideration must be taken of this fact.

Research so far undertaken has succeeded in establishing a number of important general design criteria, defining these as appropriate human centred design criteria for systems having to cope with greater changes in product markets and greater process variations. Moreover, it is also emphasised that the conditions of particular firms have to be taken into consideration. However, what is now required is an analysis that relates human centred system design to different types of firms and industries, i.e. between *general approaches* to 'human centred' design and the 'requirements' of an era of flexible specialisation or post-Fordism and adaptations to individual firm conditions. In the terms of Sorge and Streeck, this requires a systematic 'morphology' if different industrial types, as well as a more detailed knowledge of the patterning of organisational conditions by national systems of innovation and skill formation.

A third feature concerning design criteria is the need to provide a specific illustration of the overall structure of the human centred factory. Much of the previous work on human centred criteria has employed the image of a semi-autonomous production island as an illustrative scenario for discussion in the design process. What is now required is a broader scenario of the human centred factory. An image is required not only of how shop floor personnel *take on* new functions and are trained appropriately, but of how design and planning personnel *give up* such functions and are trained to do so. Some important insights have been provided into such a vision by the Data Management Group of the ESPRIT 1199 project in their report on 'A Reference Model for Human Centred CIM Data Management'[21]. A concrete vision is required of the new tasks to be taken on by such personnel, and the nature of the new collaborative and coordination tasks undertaken by representatives of design, planning and manufacture. Important choices are available about integration in individual jobs or through the establishment of new forms of teamwork, and in both cases the details of which jobs are to be expanded and how the teams are to be organised, i.e. the range of functions they take over from separate departments, representation from within and between departments, the resources and authority at their disposal etc. Such a project may draw upon the present experimentation and assessment in this area in 'design for manufacture' research and projects.[22] In addition, the insights into 'family of products'-based production islands and design insula may be extended into 'product'-based rather than 'functional'-based organisational divisions, drawing upon the substantial organisational research into this area.[23] At present human centred CIM design work in this area, carried out by groups such as those working on the ESPRIT 1217 project, has only just begun.

Design Process

In the specification of human centred design processes, the research on CAD/CAM suggests a number of areas for further research. If technological systems are seen as variable configurations undergoing a complex process of intra-firm development, human centred system designers carrying out a project within a firm should be seen as providing a relatively short term injection of system ideas and elements in an ongoing process of change. An important element of the design process is, therefore, to tailor the conduct of design projects to these intra-firm dynamics. It is, therefore, necessary to understand the 'social constitution' or 'industrial culture' of the firm.

In order to ensure the successful implementation of the project it is necessary to discern the facilitating and hindering forces to human centred initiatives within the firm. On this basis it is possible in the design process to either adapt the system design to take account of these forces or generate techniques and mobilise resources to overcome opposition. As frequently observed in job design projects, the failure of job designers to pay attention to the organisational context (e.g. systems of management control and supervision, level of employee expectations and abilities, payment system etc.) has frequently resulted in an inability to implement new systems, the confinement of the change to a small section of the workforce, or a gradual erosion of the new ways of working and a regression to traditional methods.[24]

In order to deal with these problems it is important to tailor-design processes to different organisational conditions. However, as emphasised above, this requires in the words of Sorge and Streeck, a systematic study of the population of different types of firms and industries in different countries. This is a further argument for the systematic study of industrial cultures and social constitutions as a central feature of the human centred CIM design movement. The current work carried out by FAST on national production cultures and their implications for the design and diffusion of anthropocentric systems represents an important step in addressing this issue.

In a general sense the focus on the broader character of human centred design processes should also involve an examination of the existing method of funding human centred research through the direct promotion of 'pre-competitive' technology projects. As mentioned in reports ESPRIT 534 and ESPRIT 1217, the nature and directives of the funding institutions clearly influence the type and scope of activities of the system designers. Moreover, a large number of humanisation of work initiatives in Scandinavia and Germany have involved different forms of government funding for their activities. The study of the influence and appropriateness of direct, indirect, and indirect specific inititiatives would be a useful extension of research on human centred systems design. This could draw on existing studies carried out, for example, into the nature and experiences of 'socially oriented technology policy' initiatives.[25]

Human Centred Systems and Production Paradigms

Human centred systems design is frequently presented as providing a new design 'paradigm'. This has been influenced by the original Kuhnian use of the term within the history of science, and its extension by innovation theory into the area of 'technological paradigms' and by industrial sociologists to 'production paradigms'.[26] The use of the term 'paradigm' to refer to human centred systems methods is also frequently accompanied by a sharp contrast between 'technocentric' or 'Taylorist' approaches to systems design.

In some senses semantic quibbles over the applicability of the term are not very useful, however in one sense its past use in human centred system design can be seriously misleading. If traditional systems designs are seen primarily as an 'expression' of a 'Taylorist' paradigm, then the major task of human centred researchers is to build up evidence against this paradigm, and assist its replacement by developing and showing the usefulness of an alternative paradigm. This leads directly to a political strategy that emphasises the central importance of directly participating in 'exemplary projects' designed to show the superiority of the new paradigm.

The problems of this approach are threefold. Firstly, the development of systems is, as we have seen, a consequence of a complex and negotiated process of development. To represent such systems as the direct result of a 'design' paradigm is, therefore, to misunderstand the concrete conditions responsible for their development. A factor that seriously hinders an adequate understanding of the actions required to effectively transform such systems[27]

Secondly, the extent to which a 'Taylorist' paradigm is actually adhered to by managers and systems personnel within manufacturing firms is uncertain. In the debate surrounding 'flexible specialisation', 'post-Fordism', and 'new production concepts', this has been a major point made by the critics of such theories.[28] To a degree management has frequently resisted too extreme a Tayloristic view, and the dual character of human resources as a source of potential value as well as disturbance is widely recognised. This is revealed in contemporary projects and discussions of 'neo-Fordist' or 'neo-Taylorist' systems that attempt to exploit the advantages of integrated reprogrammable technologies utilising limited forms of worker flexibility and job enlargement, without extending these to the level of worker autonomy and job enrichment envisaged by human centred systems designers.

In this situation, the over-preoccupation of critical attention on the 'inefficiencies' of the Taylorist paradigm and the 'efficiencies' of the human centred paradigm may be weakened as the 'neo-technocratic' emphasis on the objective efficiency of the new paradigm, may be regarded highly sceptically by management and systems personnel aware of the socially constructed and negotiated nature of 'skill requirements' and 'efficiency' tests. This social construction has been frequently observed in discussions of figures being 'concocted' by engineers in justifying the introduction of CAD/CAM systems.[29]

Thirdly, technological paradigms have been employed in three different ways in the literature. They have been used to refer to: objective indicators of technological trajectories resulting from established technological communities or regimes producing technological equipment or complexes; alternative historical paths of production systems development within user organisations, and as intellectual 'visions of efficient production' influencing production system development.[30] While the idea of human centred design as a new paradigm may be appropriately used in the third sense, it has been often inappropriately employed in the second sense, and been rarely applied to detailed investigations of technological paradigms in the first sense.

The latter is a significant ommission, a detailed examination of the technological paradigms governing the process control and computer industries producing CIM components and systems would provide an important basis for assessing how European technology policy could intervene to influence development in a more human centred direction.[31]

Conclusion

In conclusion, the statements made here are intended as general guidelines for extending human centred CIM design research, they do not yet contain specific proposals on how this should be undertaken. The comments should not, moreover, be taken to any way devalue present human centred design research: it has I believe performed an important role, and there is a great need for more of the same kind of research. The comments suggest, however, the need for new forms of interdisciplinary collaboration in the extention of the human centred design approach. Initiatives of the form presently being undertaken by FAST and the CAPIRN network are an important example of the kind of developments suggested here.

It has frequently been mentioned in human centred CIM and other humanisation of work projects that there was a great communication problem between the technical personnel and the social scientists, and that great efforts are needed to overcome this divide. Similar efforts, possibly no less difficult, are also needed to overcome the divisions *within* the social sciences, in particular the separations between psychologists, ergonomists and human factors engineers; industrial sociologists; innovation theorists; sociologists of technology; and technology policy analysts. If there is one clear conclusion to emerge from the observations mad in this paper, it is that in order to address the complexities of conscious socially responsible systems *redesign*, and effectively intervene in the social shaping of new technology, new forms of interdisciplinary collaboration are required.

Notes

[1] See, for example, R.U.Ayres, (1990) 'CIM Program Overview: Macro', IIASA *CIM: Revolution in Progress* Conference, Vienna, 1-4 July

[2] See P.Brodner and P. Kidd's discussions in *Anthropocentric Technology & Systems (AT&S) Newletters* 1 and 2, Brussels Februrary and June 1990

[3] Badham, R. and Wilson, S. (1990), *Computers, Design and Manufacturing: The Challenge*, Employee Participation Report, Department of Industrial Relations, AGPS, Canberra, Australia

[4] Rosenbrock, H. (1989), *Designing Human-Centred Technology: A Cross Disciplinary Project in Computer Aided Manufacturing*, Springer, London

[5] Marx, K. (1975), *Capital Vol.1*, Lawrence and Wishart, London, .p.384/5)

[6] Ibid, p.385

[7] See M.Cooley (1987), *Architect or Bee*, Abacus, London; and Susman, G. and Chase, R. (1986), 'A sociotechnical analysis of the integrated factory,' *Journal of Applied Behavioural Science*, 22:257-70

[8] In addition to the discussion carried out in the CAD/CAM report, see McLoughlin, I. and Clark, J. (1988), *Technological Change at Work*, Open University, Milton Keynes

[9] Wynne, B. (1988), 'Unruly Technology: Practical Rules, Impractical Discourse and Public Understanding', *Social Studies of Science*, 18, 147-168

[10] Kaplinsky, R. (1984), *Automation: The Technology and Society*, Longman, Harlow

[11] Sorge A. and Streeck, W. (1988), 'Industrial Relations and Technical Change: The case for an extended perspective', in R.Hyman and W.Streeck (eds.), *New Technology and Industrial Relations*, Blackwells, Oxford; and Child, J. (1987), *Organization: A Guide to the Problems and Practice*, Harper and Row, London

[12] Child, ibid.; and Bessant, J. and Grunt, M. (1985), *Management and Manufacturing Innovation in the UK and West Germany*, Gower, London

[13] Buchanan, D. and Boddy, B., (1983), *Organisations in the Computer Age: Technological Imperatives and Strategic Choice*, Gower, Aldershot

[14] Personal Interview (1990), Burkhard Shallock, Planungstechnik, Fraunhofer IPK, West Berlin

[15] Hildebrandt, E. (1989), 'The Social Constitution of Firms', Paper presented to *CAPIRN Workshop*, Santa Cruz, November

[16] Ibid., p.9

[17] Sorge, A., Hartmann, G., Warner, M., and Nicholas, I. (1983), *Microelectronics and Manpower in Manufacturing- Applications of Computer Numerical Control in Great Britain and West Germany*, Gower, Aldershot

[18] See, for example, the discussion in R.Badham and B.Schallock, (1990), 'The Human Shaping of CIM', *IEE Proceedings IT and People Conference*, Bournemouth; and Williams, K., Williams, J., and Thomas, J. (1984), *Why are the British Bad at Manufacturing?*, Routledge and Kegan Paul, London

[19] Clegg, C and Symon, G. (1989), *A Review of Human-Centred Manufacturing Technology and a Framework for its Design and Evaluation*, MRC/ESRC Social and Applied Psychology Unit, University of Sheffield, Memo No.1036

[20] See Brödner, P., (1988) 'Options for CIM: "unmanned factory" versus skill based manufacturing', *Computer Integrated Manufacturing Systems*, 1,2, 67-74; and E.Havn et.al.(1988), *A Reference Model for Human Centred CIM Data Management*, ESPRIT Project 1217 (1199), Deliverable R.16

[21] E.Havn et.al., ibid.

[22] For example, more detailed discussions of what this means in terms of manufacturing representatives in new product teams; for example, does this involve line management, production engineers, supervisors or craftsmen; do they have a veto over decisions or merely a right to input; are they in a minority or a majority compared to design personnel; is one of them the chairman of the group or not; and what decisions about new production equipment and techniques have to be referred to this group, and what control does it have over investment and planning. See the discussion in Badham, R., and Wilson, S., ibid.

[23] Child, J. (1987) ibid.,

[24] Kelly, J.E. (1982), *Scientific Management, Job Redesign and Work Performance*, Academic Press, New York

[25] See, for example, Badham, R.(1990), 'Towards a Socially Oriented Innovation Policy', Paper presented to *'Socially Oriented Technology Policy' Conference*, Institute of Advanced Studies, Vienna. Also, chapter in forthcoming book, G.Aicholzer et.al. (eds.), (1991) *Socially Oriented Technology Policy*, De Gruyter,

[26] Clark, N. (1987), 'Similarities and Differences Between Scientific and Technological Paradigms', *Futures*, Feb, 26-42; and Piore, M. and Sabel, C., (1984), *The Second Industrial Divide*, Basic Books, New York

[27] This is also something that is now increasingly recognised within the history of science as a factor influencing scientific development. Paradigms are continuously negotiated phenomena, cultural resources that are shaped and deployed in negotiations within and between scientific communities. See Badham, R. (1989), 'Metaparadigms and Better Paradigms: Thoughts on the Social Construction of Technological Systems', *STS Seminar Series*, University of New South Wales, Sydney. As discussed by Norman Clark, (ibid.,) this is a phenomenon that is made even more unstable and complex in the case of technological paradigms.

[28] Badham, R. and Mathews, J. (1989), 'The New Production Systems Debate', *Labour and Industry*, 2,2, 194-247

[29] See, Badham, R. (1989), 'CAD/CAM,. Work Organisation and the Integrated Factory', *IEEE Transactions on Engineering Management*, Special Issue: Social Impact of CAD, August

[30] Badham and Mathews, ibid.,

[31] An indication of the form that such an analysis could take is provided by present work carried out by Dr Hagedoorn at the MERIT Institute in Maastricht, Netherlands.

Human Centredness: A 21st Century Paradigm for Industrial Cultures

Karamjit S Gill

Crises of the Post Modern Industrial Society

The cultures of the 'assembly-line' management and the 'machine-centred' technology have dominated the shaping of the modern industrial cultures for over half a Century. The technology-centred modern industrial cultures are now facing challenges of post-modernity arising from the limitations of the dominant paradigm of mechanism, and its impact on the shaping of information society and the world of work.

Crisis of the Mechanistic Paradigm

We find that established paradigms of science and technology, the Newtonian-Cartesian 'mechanistic paradigm' and the Tayloristic 'management science' paradigm are being increasingly challenged as regards their relevance to the design of new technological systems. These challenges arise out of the narrowness of these paradigms which limit humans capability to shape work, technology and environment consistent with the requirements and aspiration of the industrial cultures and social systems of the 21st century Europe. In the world market of 'competitive displacement', there is a serious concern about the 'management science paradigm', and its' ability to develop a high quality flexible and adaptable manufacturing base without questioning production structures which merely attempt to replace human work and skills. Alternative production systems are gaining attention which aiming to combine unique human abilities and skill with the high performance of machines.

The mechanistic paradigm views the human mind as a symbol processing machine, thus leading to the assumption on the part of cognitive psychologists that it is possible to explain human behaviour or at least produce human behaviour through designing intelligent machines. This computer metaphor is used to propagate the idea that 'humans making mistakes and being unreliable can be replaced by superior machine systems'. The largest potential of this rationalisation is seen to lie in designing the automated office. Although the automation of complex information processing tasks such as drafting, planning, inventory control and scheduling may have reduced costs, the rationalisation does not seem to have made any fundamental improvement to the office working situation.

There are problems concerned with the integration of intelligent systems with complex human working environments (which remain either misunderstood or unresolved). Systems are designed to the rationalist criteria of correctness, consistency and reliability. These systems are based upon the assumption that the tacit dimension of the human knowledge of complex working environments can be explicated and formalised for the design of such systems. However, marginalising the subjective human element leads to inconsistency and unreliability because of the inflexibility of the system to deal with new cases which require human intuition. This leads to the increasingly recognised problem of maintenance of complex systems.

The ironic consequence is that management is becoming dependent upon incomprehensible and unreliable production and office systems. The alternative human-centred paradigm focuses on combining of the contrasting capabilities of human and machines productively, with the aim of enhancing the productive and creative potential of humans rather than imitating their abilities, or reducing the abilities to the functionality of the machine; or even to remove them from the productive process altogether.

The mechanistic paradigm has been so successful in its paradigmatic technological applications that it became the ideal prototype for nearly all scientific thinking. The post-war period has seen the acceptance of the Tayloristic paradigm of the division of work and the division of time as a basis for modelling industrial and management systems. There is, however, increasing recognition that the success of a production process depends upon the degree to which the knowledge and skill of the human is enhanced when working with the technological systems rather than on the degree to which human work has been displaced by the machine artifacts. The division of academic disciplines as well as the boundaries of professions are products of the mechanistic paradigm. Now, as we move into the era of quality revolution of high quality and small batch production requiring a highly skilled, flexible and adaptable work force, there is a need for a new paradigm which recognises the environmental and social concerns of the world of working life, and builds upon the strengths of human knowledge, experience, skill, creativity and ingenuity. Human centredness may become a new paradigm for the 21st Century.

Crisis of the Information Society

In recent years an historical transition has taken place, from activity involving the creation and manipulation of artifacts to activity based on something called 'information', and that this activity takes place in an 'information society'. One of the central themes of the information society is *knowledge is power*, where knowledge is treated as a product that can be produced from human raw material, and as a commodity that can be marketed in the competitive economic world. This view of knowledge separates knowledge from the social and cultural contexts in which it is constructed, interpreted and maintained through its practical use in context .

This 'information society' view of knowledge is derived from the Platonic view of Knowledge. Plato attributed to Socrates the belief that true 'knowledge' could be stated precisely. Anything which could not be so formulated was not knowledge but merely belief. Modern day Platonists, such as Bell, treat knowledge as property, as a commodity, which Cooley says, puts most of us firmly in our place; we are *consumers of knowledge*, a commodity produced by others: the 'experts'.

The message of 'Information society' is that Knowledge is something which others have and to which I have access, second-hand. The risk is that, as time goes by, more and more sections of the community will lose the capacity to appreciate craftsmanship and goods of quality as aspects of current material culture. Having de-skilled the producers, IT may lead to the de-skilling of consumers too.

The 'information society' cliche, defectively focuses attention onto technology and away from fundamental issues such as the political implications of the new information and communication technologies. The 'default' view of the IT society rests on the notion of information as a traded commodity, supporting new sectors of activity within a simple market economy. This view has the disadvantages of defining citizenship in terms of the economic relationship to the production and consumption of information.

The alternative human-centred view of knowledge is echoed by Umberto Eco, who talks about 'cultural patrimony' of knowledge; knowledge is not simply created and owned by a tiny minority of licensed sages, but that much of human creativity depends on deeply embedded cultural knowledge which cannot be easily explicitly stated. Practical Knowledge is acquired through doing or 'attending to things', Common sense knowledge is acquired through learning by doing, i.e. experiential knowledge.

The need for an alternative paradigm to that of the 'Platonic' paradigm argues that the technological changes taking place in the era of information society are much more complex and less predictable than can be encompassed by the notion of the 'information society' based upon a view of knowledge as 'strategic resource', as subscribed by Bell. From the technological perspective, the most crucial development is in the field of human knowledge. The field of knowledge extraction transcends inorganic and physical nature as described in physics and chemistry. Knowledge extraction is now of a biological and organic nature, and is concerned with human consciousness, and knowledge about knowledge processes and symbolisation. The human mind is not only a creator of the world, as in the past, but it is also the object of extraction and creation. The concept of information is thus to do with both forms of knowledge, i.e. knowledge which can be formalised (objective) and implemented in the machine, and the general knowledge (tacit) which is beyond explication and mechanisation.

This concept of information, intertwining the objective and the subjective, might be described as a shift from the 'paradigm of separation' to the 'paradigm of unification'. According to this perspective, the concept of information may be used as a vehicle for crossing the borders between domains which have been separated by existing scientific traditions, i.e domains such as the organic and the inorganic, energy and information, biological processes and psychological processes, relationships between cognitive biology, psychology, and the creative cognitive potentials and human artifacts.

Paradigm Crisis of Work and Technology

Three technological perspectives, 'technique-oriented approach', 'socio-technical approach' and 'human centred approach' reflect different assumptions of human and machine relationships. The technique-oriented approach, according to Nurminen places machines or technically mediated communication before human beings and direct personal communication. The term 'socio-technical approach' seeks a point of equilibrium between the machine and the human. The term 'human centred approach' places human beings and direct personal communication before machines or technical-mediated communication, but still seeks a combination from that kind of priority. The shaping concept of human-centred approach intends to cross the borders between technical and social sciences as well as between theoretical and practical knowledge through an action-based dialogue.

The Technique-oriented approach

The technique-oriented approach is modelled on the mechanistic model of the universe. In this universe time is unidimensional, flowing from the past through the present to the future. that is, strictly deterministic. Human beings are viewed as biological machines. During the nineteenth century. The technique-oriented approach has provided the foundations for developing organisational structures and management of workers as if they were machines. In 1832, Charles Babbage, the inventor of a mathematical computer, advocated a scientific approach to organisations. It was not until the twentieth century that the mechanical approaches were synthesised in a more systematic way. The German sociologist Max Weber observed the parallels between the mechanisation of industry and the emerging bureaucratisation of organisations based upon precision, regularity, speed and efficiency through the establishment of fixed division of tasks, hierarchical structure and the detailed rules of all activities. The parallel with machines was clear, and deliberate.

The functional weakness of this approach lies in the failure to take sufficient account of the reality of work organisation and the educational needs of the end-users. The theoretical weakness of this approach lies in the inherent weakness of a positivistic epistemology of practice which builds upon three dichotomies, the separation of means and ends, the separation of theory and practice, and the separation of doing and knowing.

The Sociotechnical Approach

The concept of the sociotechnical approach was developed by the London Tavistock Institute of Human Relations in the 1950s. In this approach, the technical and the social systems of an organisation are considered more or less on equal terms. Initially its area of interest was industrial work: how the technical system, usually a manufacturing or production process, could be designed in such a way as to be socially acceptable and correspond to human needs.

The approach was a response to earlier mechanistic work-oriented approaches such as Taylorism (Taylor 1911) which saw the organisation of work as a technical problem. Much of the work-oriented research strategies of the late 1920s and 1930s were aimed to 'repair' the social and productive inefficiencies of this technical-oriented perspective. The idea of stimulating the productivity through the satisfaction of the social needs of workers became a focus of attention of a number of organisational psychologists, and different modifications of bureaucratic structures began to emerge.

During the 1960s and 1970s, the ideas of the socio-technical approach became attractive to management in an attempt to increases productivity and reduce turnover without any additional costs. The experiments of Volvo Kalmar in the 1970s and of Volvo Uddevalla in the 1980s are well known examples stressing work satisfaction and the health and safety aspects of technical functionality and the quality of products.

As compared to the technical-oriented approach, the Sociotechnical approach has much more positive concept of the human being. One of most valuable principles of the socio-technical approach is the centrality of participation. The human is perceived as a member of a group, an active and a responsible member doing 'a good day's work'. An informal communication network is considered just as important as the official one. There is at least some scope for end-user collaboration and democratic participation of the work force.

However, the approach has in principle several limitations. The organisation is treated as two separate systems, technical and social, in equilibrium of some kind. This means that the technical system is, by and large, taken for granted, and the equilibrium is achieved through making the social relations adaptable to existing technical devices, or dominated by management decisions. Thus the scope for radical changes in human-organisation relations is rather limited from the very beginning. Though the Sociotechnical approach had in many ways changed the perception of the worker in a more positive direction, the approach did not really understand the importance of subjectivity bounded knowledge in the work process.

Human Centredness: A New Direction for Industrial Society

Human Centredness is a new tradition of human activity, which places human need, skill, creativity and potentiality at the centre of the activities of human organisations and the design of technological systems. It is an emancipatory tradition which is rooted in the diverse cultural, scientific, and philosophical traditions of Europe. The tradition originated in Britain in the 1970s an as alternative to the Tayloristic approach to production and the industrial rationalisation due to rapid advances in microelectronics. Since 1970s, It has influenced the development of culturally-oriented traditions in Europe, for example the 'humanisation of technology' in Germany, and the 'democracy in participation' in Scandinavia. These developments are now providing the creation of a new European tradition of 'Anthropocentric Systems' which extends the industrial and manufacturing aspects of human centred approaches to cultural contexts.

The concept of 'anticipatory democracy' and the ideals of 'cultural and industrial renaissance' provide a human-centred framework for this emerging new tradition. Within the context of the increasingly interdependent world of diverse cultural, industrial and political traditions, and social aspirations, human centredness emphasises that technology development is a culturally-oriented process. It is based on the concept of the 'valorisation of differences', and rejects the notion of the 'one best way'. This approach supports the design and development of completely open ended advanced technologies, that is to say those which leave room for organisation, for very diverse use, and take into account the relevant social and industrial perspectives. The human-centred tradition thus provides an alternative to the dominant 'Tayloristic' management science tradition and the Cartesian approach of Western Science and places an emphasis on the adverse environmental effects of the technology dominating nature, to ensure that we do not end up with technology dominating our lives and the future of humanity.

Human centredness builds on the idea of human-machine symbiosis and regards the social and cultural shaping of technology as central to the design and use of technological systems. It emphasises the interdependence between subjective and objective knowledge by placing practice, skill and dialogue at the centre of designing human-machine collaborative systems.

Human centredness rejects the mechanistic paradigm of technology development and provides a powerful alternative philosophy for the design and use of systems. It rejects the idea of the 'one-best way', 'one culture' and 'sameness' of scientific ideas. Its alternative focus on diversity provides a motivation for looking at the structures which support innovation and reflect and enhance cultural diversity. This focus of research recognises the need for the active participation of citizens in shaping science and technology.

It is important to emphasise that human centredness is not an anti-technology or anti-science but a tradition which transcends the narrow mechanistic notions of science and technology (e.g. statistical control, objectivity, quantification), and crosses the boundaries of academic and working life disciplines.

Some of the main ideas of this emerging human centred debate are that:

i) Education is about the transmission of culture which values pro-active, sensitive, creative human beings;

ii) the 'Economy of scope' allows us to think of production rather than just products as allowed by the 'economy of scale';

iii) It is by being proactive and not simply reactive, and by designing tools and not machines, that we can cope with the highly complex and synchronised systems of the microelectronics age;

iv) It is by adopting the concept of 'unity through diversity' and rejecting the notions of the 'one best way', 'one culture', and 'sameness of scientific ideas' (e.g. consistency, reliability, predictability) that we can deal effectively with the issues of the globalisation of knowledge and the internationalisation of economies.

Human-Centred Tradition

The Human centred tradition has arisen out of two complementary approaches in Britain in the 1970s, 'human-machine symbiosis' (UMIST tradition) and socially useful production (LUCAS Plan). The common core of these approaches is that culture-based knowledge and actions of human beings should be reflected in a dynamic way in systems instead of being subsumed. The UMIST approach of human centredness' grew out of a CAD project which aimed at fostering human skill and cooperating with it, as an alternative to the dominant 'Tayloristic' approach which aimed at continual reduction and ultimate elimination of skill and responsibility in the worker (see Rosenbrock, 1989).

The rapid development of microelectronics in the 1970s gave rise to a wave of rationalisation in British industry. Workers at Lucas Aerospace put forward an action plan 'Lucas Workers Plan' for socially useful products. The idea of human centredness was identified with production that was compatible with social needs and was determined by 'use value' rather than just 'exchange value', and in which workers have a right to play a dual role, as producers and consumers. Highly skilled engineers, technicians and skilled workers, who lost their jobs due to rationalisation, turned their protest into constructive ideas for socially useful products, and demonstrated in a practical and direct way the creative power of their own skills and abilities. The Plan 'represented an enormous extension of consciousness' of 'ordinary people' (Cooley, 1987).

Since 1970s, both the UMIST tradition and the LUCAS Plan have influenced the development of human centred approaches in other European nations, for example the concept of the 'Tool Perspective' and 'Democracy in Participation' in Scandinavia, e.g. UTOPIA Project, the 'Humanisation of Technology and Work' in Germany , and very recently the 'The Shaping of Technology and Work' arising out of the ESPRIT Project 1217 (EC). These developments in human centredness in Europe are now contributing to the creation of a new European tradition of 'Anthropocentric Systems' (European Competitiveness in the 21st Century, FAST Report, 1989). At the heart of the new anthropocentric tradition lies the concept of diversity both in the philosophical sense and cultural sense. The centrality of this new tradition is that it should be both democratic and rooted in culture, reflecting cultural and linguistic diversities. Although human centredness is essentially a European tradition, and may not map directly onto other cultures and nations, the basic ideas of diversity, cultural and industrial renaissance, anticipatory democracy, and holistic systems could provide a common basis for culturally-based human centred traditions. Both the theory and practice of the human centred tradition are still open and are undergoing development. The attempt to maintain a mutually nutritious interplay between theoretical reflection and practical developments is central to the tradition.

Building upon the long intellectual and scientific traditions of Europe, the European research on human centredness would continue to expand upon the now established human-centred research traditions.

The European Commission has shown its commitment to the human centred developments through its support for research into the Anthropocentric Technology & Systems under the FAST programme. Human-centred research is supported by an increasing number of research institutes, university research centres and funding bodies throughout Europe. This research is linked by various European research networks, debating forums, new journals, new book series, and new postgraduate course developments. An International Institute of Society, Culture and Technology is being established in Europe with the aim of supporting international developments in human-centred systems.

Anthropocentric Debate in Europe

The debate on anthropocentricity in Europe aims to contribute to the evolution of the new socio-economic and technological horizons of Europe, as it builds its economic competitiveness and industrial culture to reflect the variety of social structures and cultures within European societies. The socio-economic horizons reflect the particular nature of the European production structure, and the problems posed by the SMEs, shifting markets, changing demands, broader product variety, environmental factors, education and demographic development. Technological horizons are concerned with the inflexibility and vulnerability of complex technological systems.

Some of the main ideas of the anthropocentric debate are Europe's strength through cultural diversity; sharing of human goals through the use of knowledge and technology; science and technology policy options for the enhancement of human life; education and learning as a social shaping process; social and cultural shaping of work and technology; technology as a tool and not a machine; environmental issues and demographic shifts; economies of scope rather than economies of scale.

The debate is based on the premise that 'the long term competitiveness of the European nations, economies and firms will deeply depend upon our ability for an intelligent combination of proper 'government' of the relationships of people, the technology and the organisation'. The FAST Report (Cooley, 1989) argues that since the European manufacturing base is largely composed of SMEs, and is characterised by a highly skilled and flexible work force, its future strength will depend upon the development of anthropocentric systems which build upon the skill, ingenuity and expertise of the working people. The belief is that the regional and cultural variety of Europe will provide the basis for addressing diversified markets and responding to the demand of product variety.

The anthropocentric approach- a new paradigm of a symbiosis between work and technology - would provide a basis for building the 'future-oriented European production culture'. The overiding purposes of this symbiosis will be to develop human-centred technologies which optimise performance, increase production capability, and protect the economic competitiveness of Europe. To meet these demands, the human centred technology must be integrated with, and complemented by, an equally flexible organisation of work.

The FAST debate on anthropocentric technologies is concerned with four areas:

1. Human work in advanced technological environments

2. Forms of technology which supports SMEs and regional development

3. Environment, energy resources and new materials

4. Educational aspects and human resource development

The focus on 'Human work in advanced technological environments' emphasises the scope of anthropocentric technologies which link the skill and ingenuity of humans with advanced and appropriate forms of technology in a true symbiosis of work and technology. It is deeply concerned about the indiscriminate replacement of humans by machines and warns about the lack of robustness and flexibility of machine dependent systems and their vulnerability to uncertainties and disturbances. It rejects the concept of total automation as providing an optimum solution in the European contexts and proposes that anthropocentric technologies are most suited to gain European competitive advantage, because they would enable Europe to concentrate on 'flexible specialisation' building on its traditions of flexible production systems and skilled work force.

The second focus on technology and the SMEs argues that anthropocentric technology is most appropriate for regional development and would provide economies of 'scope' for peripheries of Europe. Distributed systems would enable small towns and villages to retain and build upon their skills and industrial heritage. Communication networks together with language exchange centres would preserve local culture and assist in the business of SMEs.

The third focus, 'Environments, energy resources and new materials' views the development of new materials in the wider context of environment, energy, health and safety and the forms of skill and means of production available to communities. On a global scale, it is concerned with the issue of resource depletion and environmental changes brought about by human activities.

The fourth focus, 'Education aspects and human resource development' emphasises that a new dimension of the 21st Century will be demographic change in society which will necessitate new mechanisms and learning technologies to enable people to learn and update their skills throughout their lives. It proposes holistic forms of education for Europe, which are about the transmission of culture, and which value pro-active, sensitive, creative human beings.

Designing Human-Centred Technology

As we move into the so-called 'post-modern industrial society, human skill, ingenuity, knowledge and expertise are increasingly seen as key resource for industry and commerce. The trend has accelerated, as the centre of economic activity has shifted from manufacturing towards the manipulation and exploitation of information. So called 'Human Resource Development' is now a primary activity for forward looking organisations, and is a major strategic concern for the EC.

Recent developments of management and production practices in manufacturing industry have also tended to emphasise rather than restrict the human role. The design of production technology is becoming increasingly oriented towards optimised integration of the special capabilities of machines and humans. The developments are most marked in countries such as Germany and in Scandinavia, where there has been a move away from the deskilling and process fragmentation associated with 'Tayloristic' management ('control revolution') towards what has been described as a 'quality revolution', representing a shift of emphasis from control of individual human actions towards a socialised commitment to high quality work. Growing integration of advanced technology into production processes, management structures and decision making is gradually changing the nature of work and working environments and, thereby the nature of skills. For example, the computerisation of production processes requires new skills, not only for workers to adapt and cope flexibly with the use of new technology, but also for them to be able to communicate effectively about their work and its environment. The integration of new technology into industry raises the question of skill acquisition to cope with the changing nature of information flows, working practices and human communication. Advanced industries are beginning to favour, for both humane and utilitarian reasons, a move towards human-centred manufacturing systems.

The whole thrust of development is posing interesting problems for designers of computer-based systems. In order to design computer based systems which are appropriately integrated into organisations, it is necessary to understand the implications of computer artifacts for organisations, to recognise expertise as growing out of working life, and to understand its nature and functions within the context of a particular enterprise. This implies a different approach to systems design than the dominant machine-centred approach. Although there has always been a significant 'Human Factors' element in systems design, the strongest influences in this field have been the research traditions of ergonomics and individual psychology.

Much, if not most, contemporary 'Human-Computer Interaction' (HCI) research provides insight into the limitation of the mechanistic paradigm of technology design. The HCI research focuses on the interaction between the individual and the computer, and takes a limited view of the social contexts within which the interaction occurs. Although HCI research can be very influential in the evolution of modern interface tools, its attention to hardware construction and software engineering has led to an implicit tendency to perceive the human user as part of the systems configuration, rather than as an active member of a wider human environment. An uncritical acceptance of this situation as the norm in the field has meant that work-related activities and group practice has not been accounted for. The result is that where systems have been implemented in the work environment they have tended to become a barrier to communications rather than a facilitator for the coordination of work.

The interface between humans and computer systems, the domain of HCI research, is the potentially fertile ground for multi-disciplinary study. But despite nominal attention to more general issues, HCI has tended to be rather narrowly focussed, with particular emphasis on individual psychology and on the immediate interface between the user and the machine. This has, of course, led to the development of a range of 'user-friendly' software and hardware, but it has contributed little to our knowledge of the complex social processes within which interactions between individual people and the computer are embedded. The instrumental orientation of the HCI mainstream, as well as its explicit focus on the design of machine-oriented artifacts is dominant in Britain

From Human Factors to User-Centred Design

There is, however, a growing recognition, among many HCI researchers, of the limitation of this 'mono-user' approach and the narrowness of its application to real life situations because of its roots in the research laboratory. There is a need for bridging the 'ecological gap' that exists between the world of the research laboratory and the real world. The focus now is on the concept of effective usability in the design of artifacts and their evaluation in real life working environments. Currently, in Britain and the USA, there is a shift from human factors and 'user modelling' concepts to 'user-centred' design of HCI.

The focus of user-centred design is on how to develop technology to facilitate group working. A constituent of the frameworks for this focus is that cognition is distributed - embodied not just in the heads of the individuals but, in a fundamental way, embodied more broadly in the natural environment, in numerous tools and artifacts, in the relationships amongst people, and in the organisations, institutions and culture of the various societies. This view of cognition represents a shift away from the laboratory study of cognition which strips away the supporting context of the everyday world in an effort to study 'pure' internal processes. Technology, therefore, is a cognitive artifact. Artifacts serve the purpose of coordination, communication, information representation, storage and retrieval i.e., core cognitive capabilities. Understanding these core capabilities is the key to user-centred design for co-operative working. A standard Computer supported cooperative working (CSCW) view of interaction is of people interacting with technology as an organised group. The user-centred work extends this notion of interaction to a study of human-human interaction and thereby understands the role of mediating technologies. The emphasis is therefore on crucial aspects such as human relationships. It is important to move on from the idea of technology as an instrumental tool to that of technology as a 'support'. This is just one relation between technology, people and work.

From User-Centred to User-Involved Design

In some parts of Europe, especially Scandinavia, there is a further shift from user-centred to user-involved design, which is based on the view that the user is not just an object of study but is an active agent within the design process itself. This involvement of users in design is seen not only as a means of promoting democratisation in the organisation change process but also as a key step to ensure that the resulting computer system adequately meets the needs of the user. It is argued that users need to have experience of being in the future use situation, in order to be able to comment on the proposed computer systems, and affect their design. User and designer can participate in the cooperative design process through mock-ups or prototypes. This perspective takes the work as embedded in the work process, and attempts to support workers through providing them with skill-enhancing computerised tools. It emphasises that the essence of the concept of cooperative design lies in respecting users skills and promoting the democratisation of all phases of design and work practice, and not in the tools or their use per se.

The challenge to HCI is to develop human-centred approaches for designing artifacts which can not only be introduced in current practice but can also enable future users to reshape their practice and skills. One such approach is the 'human activity framework' which recognises that the use of an artifact is part of social activity, and the design of computer artifacts means that we design new conditions for collective activity, e.g. new division of labour, and alternative ways of coordination, control and communication. The future use situation is the origin for design, and we undertake design with this in mind. Use, as a process of learning, is a prerequisite to design. Through use, new needs arise, either as a result of changing conditions of work or as recognition of problems with the present artifacts. To design future use activity in mind also means to start out from the present practice(s) of future users. Cooperative design is a meeting place for many different practices, for sharing experiences and learning new skills. Viewed from this perspective the design is a learning process, for individuals as well as groups, about each others practices and language games of the practices. This design approach requires that designers and users must be prepared to acknowledge each others competence and to realise that effort must be made by both parties to develop a mutually agreed upon vocabulary of concepts which can be shared across different groups in a design project. This Human activity based approach provides an alternative paradigm to the traditional individualistic cognitive science and the user-centred approaches. The Human Activity approach for HCI design forms an important aspect of the emerging new human centred tradition in computer artifact design in Europe.

The major area of HCI which is being shaped by human-centred approaches is the area of computer supported cooperative working. The main influence in this new direction comes from the Scandinavian tradition of 'democracy and participation' which originated with the UTOPIA project . UTOPIA has developed two alternative conceptions of the computer. One is the 'tool' perspective and the other is 'design-by-doing' perspective. The basic idea of the tool perspective is to view the computer artifact as a skill enhancing tool for skilled workers to produce quality products. This means taking the labour process as a starting point, and seeing the worker as someone who is in control of a set of computer based tools. This process of designing computer-based tools means that users should be involved in the design process.

This is a radical shift from the earlier systems design processes. The tool approach stresses the crucial role of the tacit knowledge possessed by people as opposed to just focussing on their theoretical knowledge. One of the crucial epistemological issues for systems design is what Wittgenstein called 'language-games'. Language games express the practices of both the users and the designers, the users are experienced in the language games of their work or use situations, and the designers are experienced in the language games of design. One of the central concepts of the UTOPIA Project is the 'use model' which draws together the relation between the user interface of the computer artifact, the users' professional language and competence, and the design situation.

The Tool Perspective

In the tool perspective, the essential parts of the user's skills which are relevant when using a 'tool' are 'tacit', and as such they neither can nor should be made explicit. The idea of the tools has been influenced by the way design of tools takes place within the traditional crafts. The idea is that new tools should be designed as an extension of the traditional practical understanding of tools and materials used within a craft or profession. It means that experienced users and designers should be involved in the design of new tools. In order to ensure that users and designers use a common language for communication and enrich their skills by learning from each other, a common use model should be created.

The use model relates the use of the computer based tools to the use of the users' traditional tools. The power of the tool as a design perspective has been demonstrated by the work of the UTOPIA project (e.g. Ehn 1988). Here, in addition to skill maintenance and development, future scenarios were created, new ideas explored, and possible design assessed in collaboration with graphic workers of a newspaper. The computer was there to support text and image creation by graphics workers. Hellman (1989) emphasises that the value of the tool lies in its metaphorical use and discusses two aspects of the tool metaphor, one that supports the design of information systems and one that can be used to improve and evaluate the quality of use. Thus both perspectives can be perceived as aiming at user-centred computing environments. In this sense, *computer as a tool* is a practical metaphor.

Participatory Design

Participatory design approaches build upon the 'tool approach' and extend the boundary of the 'user-involved' design to the design of technology supported environments. The participant in the design process is recognised as both the 'user' and 'producer' of knowledge, and the artefact is designed as a part of the overall environment. The 'environmental' perspective of participatory approaches allows for the social and cultural shaping of not only the 'tool' but also the environment in which the tool is to be used. Two projects, 'PAROSI' and 'The ELECTRONIC SKECTCH PAD' are summarised here to illustrate the use of the participatory design approaches.

The 'PAROSI' Project

The project 'Parosi' (Gill, 1989) evolved a participatory approach for human-centred systems design for complex socio-cultural and socio-economic domains. 'Parosi' was concerned with the development of a multimedia computer based tool for collaborative learning. As part of the Parosi project, a pilot 'knowledge based system for diet planning and health was developed'. The project was part of the EEC Social Fund sponsored programme on Basic Education in Numeracy and New Technology, 1983-85.

The project team involved women students, language teachers, computer scientists, experts in diet planning (health visitors) and, social mediators (voluntary researchers) from the local community. The primary aim of the 'pilot' was to develop a computer-based tool for use by students to enhance their social and communication skills. However, as the design process progressed, the aim was expanded to include the use of the tool by the design team to exchange knowledge, by experts to enhance their social and cultural competence, and by voluntary researchers to enhance their skills as social mediators. The learning environment chosen for the project was continuing education for adults, and the domain chosen for the pilot project was diet planning. The students were mainly women from diverse cultures with varied levels of language and communication skills and life experiences. The students were experts in their own right and brought to the project their own experiences of cooking, food selection, and their knowledge about diet planning.

Throughout the design process, the participants collaborated in gaining insight into the language games required for knowledge transfer both through the sharing of cultural experiences and the exchange of domain knowledge. The learning environment provided a complex social environment in which participants were able to act as learners, as teachers, as experts and as social mediators at various levels of knowledge exchange. Language games is the expression of the practices of both the users and the designers, the users are experienced in the language games of their work or use situations, and the designers are experienced in the language games of the design.

Systems designers and users in the design teams have to learn each others language games in order to be involved in the cooperative design process. This idea was fundamental in the Parosi project. The students were learning about the language games of the experts, and of the social mediators, teachers, computer scientists, as well as the language games of other students in order to participate in the knowledge exchange during the design process. Social mediators were trying to understand the language games of the domain experts and teachers in order to mediate in the knowledge transfer between students and domain experts. Software engineers participated in the teaching process so as to gain insight into the learning process, knowledge of the project domain, and the language games of other participants.

In the Parosi project, the learning environment of diet planning provided the use model for the design process. All members of the teams were involved in the learning process, and the knowledge and language skills acquired by the teams was not of benefit just for the project duration but was for the benefit of future use in other learning and real life practical situations in the complex social domains. The design-by-participation approach of the Parosi project shows the great importance of the nature of dialogue between researchers, who are also designers, and users for affecting the design process. Emphasis is placed on the knowledge, experiences, intuition, expectations and cultural backgrounds of users. Parosi underwent the process of cooperation with diversity of interests as the project involved a multidisciplinary team of designers and users.

In the Parosi project, the design of the computer artifact (a knowledge based system) required prototyping as part of the design process. The prototyping involved students and designers in the activity of cooking and diet planning, as well as understanding of various cultural perceptions and practices of diet and health. In this sense, Parosi can be compared again to the Utopia project in its focus on action research and design-by-doing method which enables the users to express their know-how in action, and involves both the researcher and the user as active contributors to problem understanding and problem solving. In the case of Parosi, action research was to do with the researchers and users involved in learning about each others cultural and social worlds, and building the computer based artifact collaboratively.

Parosi shows that creative learning is part of the dialogical process. The social aspect of expertise and learning is central. That is, the discussion between the women students and designers in social settings was as important as in the setting of the group meetings with designers. The informal and formal aspects of communication structures are as important as communication between each other for the participatory design process. The choice of food for diet planning which is an area of common interest and concern allows for the development of a common language. This language also has meaning because of its use situation. The language of cooking expresses the practice. The participatory learning approach is an action oriented and collaborative activity requiring cooperative working. It is significant that knowledge lies in the social life of the group, not just in an individual's head.

The knowledge acquisition process in Parosi was in itself a participatory learning process and the knowledge representation structures were dependent on the nature of the knowledge dynamics. Part of the dynamic function of a human centred design process of a system is the idea of a collective knowledge base which is developed on line for the sharing of the knowledge amongst all the participants. In the case of the Parosi project the objectives of the design of the technology were being achieved in the design process itself, i.e. that of the distribution and transfer of knowledge in the social domain - the design process was intrinsically linked to the use of the end system.

Shaping the philosophy of Parosi involved the learners and researchers in crossing their own language, cultural, professional and academic boundaries in order to exchange knowledge and experiences. This exchange took place both in informal social settings as well as formal learning settings as part of the learning process. These settings enabled the participants to act as learners, users, teachers, and designers at different moments, thus learning about each other's language games.

The Tool perspective in the Parosi sense meant that the knowledge based system reflected the existing practices of the designers and users while aiming to enhance the skills and competence of both the learners and the designers for the future use of these tools in new learning environments. The construction of the knowledge based system for diet, planning and health was based on the cultural practices of cooking and the perceptions of health care on the part of learners, and methods of diet planning on the part of the professionals e.g., health visitors.

The Parosi philosophy of participatory design and cultural diversity provides a developmental framework for research into transcultural knowledge transfer and cultural exchange. The essence of the project was to regard the design process as part of the learning process which focussed on language development and skill enhancement through cultural exchange in the formal settings of the school and informal social settings of the playground and community activities of learners. The Parosi approach is also reflected by the European Commission project, MEDICA which is part of the European Commission AIM programme Exploratory Action. (Smith D, 1989).

One of the central ideas of the emerging human centred research is the acceptance of diversity, whether this exists within the multidisciplinary design team or within the richness in the diversity of culture, values perspectives, experiences and skills of the participants. In the Parosi project, women learners brought their knowledge on food and processes of cooking, often related to health care from their own cultural perspectives, which they had to explain to each other and to the 'designers' who also acted as learners. The strength of Parosi was its philosophy in seeing the learning takes place as a process of cultural exchange rather than just as a transfer of knowledge and skill. In this sense, the Parosi philosophy reflected the future anthropocentric approach to education in Europe argued for by Cooley (FAST report, 1989).

The Electronic Sketch Pad' Project

The 'electronic sketch pad' project was the Danish Group's contribution to the participatory design approach of the ESPRIT project on Human-centred CIM Systems. Lœssøe and Rasmussen (1989) give a detailed description of the sketch pad development. The Danish group developed a demonstration model of a new CAD input media: an 'electronic sketch pad'.

CAD systems built for use by industrial designers follow the strategy of subdividing and codifying the designer's knowledge and subsequently reducing the designer's activity to a series of choices in a formalised standardised system of possible actions.

Human centredness, however, follows the belief that the work-culture based experience and action of designers are being reflected and prioritised in a dynamic relation to new technological possibilities instead of being subsumed or made obsolete. The latter being the philosophy of scientific rationalism behind CAD systems. This advocates increasing divisions and intensification of design work which in the long run is inferior to and against a human-centred strategy which believes in increasing the competence of self-organisation and innovation.

The application of human centredness to the Danish project needed a methodology which enabled the researcher to be in continuous dialogue with practical experienced industrial designers through all the R&D phases. The human centred approach requires an understanding of the industrial design process and industrial designer as elements of a work culture. This includes 'material aspects' (e.g., tools used by designers, products designed) and 'mental aspects' (e.g., how designers practice design, and how they may learn new kinds of practices).

Participation: The emergence of the idea for the Sketch Pad.

The idea emerged out of the negative experiences of using CAD systems. These included:

a) limitation of the creative activities of industrial designers in certain phases of the design process;

b) CAD systems promote isolation; induce passivity; and alienate the designer from the world of machines, equipment, and their operational details;

c) The CAD screen had a negative dissatisfied response. In comparison with the overall view at the traditional drawing board, the designers felt they were peering through a key hole.

This led to the aim of shaping new human-centred equipment. The approach followed was to combine the positive aspects of the traditional drawing board with those of the CAD system. In the first year this idea was developed in various ways. The solution for the electronic drawing board was chosen by the technical groups alone based on their expertise of existing hardware and software.

Whilst the technical group worked on the design of this board, groups of users were brought together by the social science group. The groups were initially invited to evaluate and comment upon the technical approach, and after giving a negative response, partly due to misunderstandings, were invited to explore the problem of the electronic drawing board in depth. A meeting between three representatives from the three user groups respectively, led to a suggestion by one of the representatives of improving tools of communication between designers and the pattern makers. He wanted a handy, portable electronic sketch pad to facilitate the personal communication he was used to having with colleagues and, particularly pattern makers, and wanted the ideas tested in a practical context. This shows how the idea of the sketch pad emerged from someone thinking of their needs in the practical working context and being given the opportunity to express those needs and participate in determining the design process and objectives.

The development of the sketch pad should be regarded as a step in the development of flexible user-interfaces and flexible input media for a new generation of design systems. The flexibility and human centredness of the sketch pad shows in the time taken to learn it, enhancement of learning possibilities, as a tool for dialogue, and its non-deterministic nature i.e. one may use it at any stage of the design process along with a traditional sketch pad as wished, its ability to serve as an 'idea bank' or knowledge base of sketches which can be shared etc. From the standpoint of economics, it is a small, flexible and relatively inexpensive instrument in contrast with CAD systems. It saves on time and resources in the drawing office.

The electronic sketch pad may be viewed as a modest example for improving the technical facilities supporting creative and communicative design activities. The realisation of this aim depends not only on technical functionality, but on the *social worklife* and *work culture* established in the design departments. Worklife include both explicit, regulated practices as well as non-formalised, situational and spontaneous activities and expectations. It is a complex of individual and collective actions and social relations, and forms an important part of work culture. The human-centred approach requires that product development should be organised in semi-autonomous groups with participants from all concerned departments from the outset of the development process. Rather than having a fixed division of tasks, as in the Tayloristic fashion of systems design, a multi-disciplinary team collectively possesses skills needed in the development of a product in a pattern of overlapping skills and knowledge. This kind of organisation transcends traditional hierarchical divisions between development, work preparation and production, thereby creating a more flexible structure for a direct communication network. The Danish human-centred work process presupposes vocational training and education in a close dynamic interchange with the daily work situation and subjectively bound skill improvement.

Methodological challenges

The human-centred approach implies a transcendence of traditionally bound barriers between theoretically and practically experienced knowledge. Hence user-researcher cooperation is inevitable. It is impossible to design future work places allowing for human need without openly discussing the issue with those possessing practical experiences.The initial clarification of the problem of the research project is a more open and sensitive phase in the human-centred approach than in a traditional, positivistic approach. Different interests and resources of researchers and users for cooperating often need to be discussed and modified according to commonly defined objectives and processes of dialogue.

During the dialogue process many methods may be used: experimental workshops, prototyping, organisation play, multi-metaphoric approaches. The common feature of these being the open, experimental and interactive character. Dialogue possesses opportunities of creating deeper, more reflective and more action oriented insights and /or models of innovative tools and organisational structures. In this respect, it is revealing that the subjective element in the researcher-designer dialogue was important in regarding concrete problem solutions.

Ideas based on different interests and values may be confronted in a dialectical process and thus transcend the limitations of one single perspective, as does the scientific rationalistic aim for order and analytical understanding, and move towards a more promising synthesis on a higher level of understanding and action.

Concluding Remarks

Technological cultures face challenges of vulnerability, reliability, risk and brittleness arising out of the ever increasing complexity of technological systems. Industrial cultures face challenges of economic competitiveness arising out of the dependencies on automation and mass production. These challenges are the product of science and technology which is rooted in the 'mechanistic paradigm'. There is an urgent need to recognise the appropriate boundaries of the 'mechanistic paradigm' to the needs and aspirations of the 21st Century societies.

Human centredness is increasingly being recognised as the alternative 21st Century paradigm to the 'mechanistic paradigm' which has dominated technological developments. Although the notion of human centredness has its origins in the European industrial cultures, and has its roots in the cultural diversities of European nations, the centrality of diversity allows for the emergence of diverse concepts of human centredness which are culturally and socially shaped.

References

Bannon, L J (1989) From Cognitive Science to Cooperative Design, in *Theories and Technologies of the Knowledge Society, Centre for Cultural Studies*, Aarhus University

Bell D (1980) The social framework of the information society. In Forester T (ed). *The Microelectronic Revolution*, Oxford, Blackwell

Bødker, S (1989) *Human Activity Approach to User Interface,* DIAMI PB-291, Aarhus University.

Brödner, P (1989) *In Search of the Computer-Aided Craftsman.* In *AI &Society*,Vol.3.1,

Brödner, P (1990) *Towards the Anthropocentric Factory*, International Workshop on Industrial Culture and Human-Centred Systems, Tokyo-Keizai University, Tokyo

Brödner, P (1990) *The Shape of Future Technology*, Springer-Verlag

Cooley, M (1987) *Architect or Bee?: the human price of technology*, Hogarth Press, London

Cooley, M (1989) Human-centred Systems. In *Designing Human Centred Technology: A Cross Disciplinary Project in Computer Aided Manufacture*, Springer-Veralg

Cooley, M (1989) *European Competitiveness in the 21st Century*, FAST, EEC

Cooley, M (1988) *Look, No Hands*, A Channel Four Film, 'Equinox, Series', October 1988.

Corbett, M et al (1990) *Crossing the Border*, Springer-Verlag

Ehn, P (1988) *Work-oriented Design of Computer Artifacts*, Swedish Centre for Working Life, Stockholm

Ehn, P & Kyng, M (1987) The Collective Resource Approach to Systems Design. In *Computers and Democracy - a Scandinavian Challenge*, Averbury, Aldershot

Ennals, R (1990) *AI and Human Institutions*, Springer-Verlag

Finnemann, N O (1990) Computerisation as a Means of Cultural Change. In *AI&Society*, Vol.4, No.4

Gill, K S (ed.) (1986) *Artificial Intelligence For Society*, John Wiley & Sons, Chichester

Gill, K S (1986) The Knowledge Based Machine: issues of knowledge transfer. In *Artificial Intelligence For Society*, John Wiley & Sons, Chichester

Gill, K S (1984) Crisis and creation - computers and human future. In *Artificial Intelligence: human effect*, eds. Yazdani and Narayanan, Ellis Horwood

Gill, K S (1985) *Basic Education and New Technology*, Final Report, EEC Social Fund, European Commission

Gill, K S (1988) *Expert Systems and knowledge transfer*. In Bernold, T and Hillenkamp U (eds., 1988), *Expert Systems in Production and Services,* North Holland.

Gill, K S (1988) Artificial intelligence and social action: education and training. In Göranzon B and Josefson T (eds, 1988) *Knowledge, Skill and Artificial Intelligence*, Springer-Verlag.

Gill, K S (1989) *Legal Expert Systems for the Social Domain*, International Conference on AI and Law, University of Bologna, 3-5 May, 1989

Gill, K S (1989) Reflections on Participatory Design. In *AI & Society*, Vol. 3. No. 4

Gill, K S (1990), *A Summary of Human Centred Systems Research in Europe,* NTT DATA

Gill, K S (1990) *AI and Social Citizenship: An Anthropocentric Approach*, Tenth European Meeting on Cybernetics and Systems Research, EMCSR90,Vienna, April 1990.

Gill, K S (1990) Culture, Language and Mediation. In *Artificial Intelligence, Culture and Language: on education and work*, Springer-Verlag

Gill, K S (1990) *The Computer and the Social Citizen: Technology for Participation?*, Proceedings of the International Conference on 'Computer, Man and Organisations I', Brussels, Belgium May 1990.

Gill, SP (1988) On Two AI Traditions. In *AI & Society*, Vol. 2, No.4.

Gill, S P (1990) *Dialogical Design for Knowledge Based Systems*, International Workshop on Industrial Culture and Human-Centred Systems, Tokyo-Keizai University, Tokyo, May 1990

Gill, S P (1990) *A Dialogical Framework for Participatory KBS Design*, Tenth European Meeting on Cybernetics and Systems, University of Vienna, April 17-20, 1990

Göranzon Bo & Florin Florin, (eds.) (1990) *Artificial Intelligence, Culture and Language: on education and work*, Springer-Verlag

Göranzon, B & Josefson, I (eds.) (1988), *Knowledge, Skill and Artificial Intelligence*, Springer-Verlag

Gullers, P (1990) Automation-Skill-Apprenticeship. In *Artificial Intelligence, Culture and Language: on education and work*, Springer-Verlag

Heidegger, G (1989) Human experts and expert systems: A view from shop floor. In *AI & Society,* Vol.3, No.1

Hellman, R (1989) *Approaches to User-Centred Information Systems*, University of Turku, PhD Thesis

Johannessen, K S (1990) Rule-following and Intransitive Understanding. In *Artificial Intelligence, Culture and Language: on education and work*, Springer-Verlag

Josefson, I (1990) Language and Experience. In *Artificial Intelligence, Culture and Language: on education and work*, Springer-Verlag

Læssøe J & Rasmussen L B, *The Electronic Sketch Pad- prototype observation and organizational context*, Technical University of Denmark, Building 301, Institute of Social Sciences, DK-2800 Lyngby, Denmark

Læssøe J & Rasmussen L B, *Human-centred Methods- development of computer-aided work processes*, Technical University of Denmark, Building 301, Institute of Social Sciences, DK-2800 Lyngby, Denmark

Negrotti, M (ed.) (1990) *Understanding The Artificial*, Springer-Verlag

Noble, D D (1989) Cockpit Cognition: Education, the Military, and Cognitive Engineering. In *AI & Society*, Vol.3, No.4, 1989

Nurminen, M I (1988) *People or Computers: Three ways of looking at information systems*, Studentlitteratur, Lund

Polanyi M (1967) *The Tacit Dimension*, Anchor Books, Doubleday & Company, New York

Rauner, F et al. (1988) The Social Shaping of Work and Technology, in *AI & Society*, Feb. 1988.

Rosenbrock, HH (ed.) (1989) *Designing Human Centred Technology: A Cross Disciplinary Project in Computer Aided Manufacture*, Springer-Veralg

Rosenbrock, H H (1989) The Technical Problem. In *Designing Human Centred Technology: A Cross Disciplinary Project in Computer Aided Manufacture*, Springer-Veralg

Rosenbrock, H H (1991) *Machines with a Purpose*, OUP.

Smith, D J (1986) IT, AI, and the Electronic Sabre-tooth, in *Artificial Intelligence For Society*, John Wiley & Sons, Chichester

Smith, D J (1990) *An Exploratory Action, Project Media A1037*, AIM Project

Shneiderman, B (1987) *Designing user interface: strategies for effective human-computer interaction*, Addison Wesley

Weizenbaum J (1976) *Computer Power and Human Reason- from judgement to calculation*, W.H. Freeman and Company, San Fransisco

Winograd, T & Flores, F (1986) *Understanding Computers and Cognition - a new foundation for design*, Ablex, Norwood

Young, J (1989) Human-centred knowledge based systems design. In *AI & Society*, Vol.3, No.2

The Evolution of Information in Human and Artefactual Systems

Susantha Goonatilake

Information appears as a variable in *genetic* (biological) systems, in *social* (cultural) systems, and in *artefacts* such as computers.

Information in these realms are of two broad kinds, *synchronic* and *diachronic*. Synchronic information in a device - whether it be biological, social or artefactual - helps maintain its integrity and its relationship to its environment. Synchronic information often exists as information that circulates within and without the device; the information circulating being often of a negative feed back kind.

Diachronic information flows 'historically', over long time spans from device to device, from generation to generation. As this information flows it has particular characteristics. The present paper describes these characteristics. The description of these characteristics are a highly simplified summary of a book by the author now in print; *The Evolution of Information: Lineages in Gene, Culture and Artefact* (Pinter Publishers, London 1990).

The purpose in presenting these perspectives here is the hope that as information is a common currency in the two realms of the artefact and the social, a theoretical relationship between the two will help identify the interactions between the two. Specially if the aim is to develop 'human centred' systems, that is systems from the view point of one information lineage, namely the social/cultural, such a perspective can be very useful.

The position taken in the paper is that three information domains exist in the genetic, cultural and the exosomatic, all three of which can be examined from the one broad perspective of an interaction with the environment resulting in a common evolutionary scheme. 'Information' is used here as a common element that persists as different lineages through time.

These information lineages exist from the beginning of life on earth. The first lineage to appear is the genetic and the evolution of genetic information from pre-biotic times to the present results in the conventional Darwinian tree of evolution. As a phylogenetic process, generally, information in genes grows in magnitude and complexity giving rise to a variety of responses to the environment.

Holzmuller (1984) has described in considerable detail the increases in information content of organisms as evolution proceeds. As these qualitative and quantitative changes in information takes place, a greater and more rapid adaptation to the environment takes place (Dobshansky 1972).

Yet, the responses of the genetic system to changes in the environment are slow, changes in genes corresponding to environmental changes is counted in generations. More rapid changes in the biological sphere occur in the hormonal and neural systems. Handling of time is counted in the hormonal system in minutes and in the neural system it is speeded up to seconds or fractions of a second (Jantsch 1980 p. 14).

Because of the neural system, there is in some animals a second line of information transfer, a transfer occurring socially from generation to generation. Social behaviour that gives rise to learning and transmission of knowledge from generation to generation occurs only in more advanced organisms. Lumsden and Wilson (1981) has classified learning in different animals in a phylogenetic tree to: (a) those that have no learning abilities or learning without imitation (the 'acultural'), (b) those that have some learning abilities including imitation (the 'proto-cultural 1'), (c) to those that have imitation abilities together with the teaching of others ('proto-cultural 2') and lastly, (d) to those that have all the three qualities of imitation, teaching and reification (the 'eucultural')

The acultural groups, which includes all invertebrates and cold blooded vertebrates, constitute the large majority of species with an order of magnitude of a million species or so. The proto-cultural group 1, the next in line, has roughly 8600 species of birds and 3200 species of mammals. The proto-cultural group 2, has a much smaller group of seven species of wolves and dogs, a single species of the African wild dog, a species of wild dog, species of lion, the two species of elephants and the eleven species of anthropoid apes. The only eucultural species is man (1981 p. 4). If one takes the animal kingdom as a whole, the facility for culture is reflected in a descending number of species.

The social groupings and structures that some animals form are dependent either on genetically endowed or socially acquired behaviours or a combination of both. The patterns of social organisation that depends on learning provides for a transmission line of non-genetic or cultural information. Cultural information flows from parent to offspring, from role incumbent to role incumbent, forming flow lines paralleling the genetic; primates having perhaps the most developed cultural lines. These lines of cultural information have autonomies of their own. These cultural lines, whether they be say, of status within the organisation or a mode of food preparation, are partly contingent on environmental factors. But once learnt and if they have survival value, they tend to persist.

Patterns of social structures can be seen to be evolving through time and rearranging themselves at least partially, according to environmental changes. The bundles of flow lines thus correspond to the outlines of the social structure of animal societies. These bundles of information could be genetic or cultural or both, depending on which transmission line of information is predominant. Cultural information is the most flexible, and its contents can be changed and flow lines rearranged more rapidly than the genetic information. For animals that can process cultural information adequately, it is an important adaptive and survival tool.

The cultural processes in animal societies that we have outlined here intensify in human societies. It is in this most cultural of all animals, the human, that culture has the greatest survival value. Cultural information is transmitted according to particular patterns corresponding to the attendant human social structures.

These social structures which pattern cultural transmission, have not been static. They have changed through the approximately 3 million year period of the existence of human like creatures on earth. Corresponding to the particular historical development of these social structures, cultural flow lines arose and took particular patterns and forms. To map these patterned cultural flow lines, one has to have some overview of the changes in social structure that occurred through human history. Thus one can detect some general changes towards increased complexity and differentiation of social structures, accompanying the changes in the modes of intervening with the environment.

Human historical evolution consists of a series of interventions with the environment. As the 'environment' we should include not only the physical environment, but also the social one - other humans. The interventions with the physical environment is done through tools and technology combined with human organisation constituting a force of production. There are rough correspondences between a given technology and the social system. Yet, the social environment as well as the social organisation of production influences deeply the nature of these correspondences, so that it is the correspondence rarely of one-to-one.

Human societies differentiate themselves generally according to their modes of intervening with their physical and social environments. Their social structures are deeply conditioned by their influences.

Human Social Structures as Skeletons for Cultural Transmission

If one were to recapitulate the approximately 3 million years of human evolution certain social structural features emerge immediately. At the very earliest period, hunter-gatherers predominate, this form of socio-economic grouping existing up to about 10,000 years or so ago. During this period the tool kits gradually change from the paleolithic to the upper paleolithic to the neolithic and in the process increasingly become more elaborate.

After the Neolithic, hunter-gatherers are gradually replaced by agriculturists and hunter gatherers increasingly become a marginal population. With the Industrial Revolution, agriculture itself becomes less important and the percentage of the population engaged in agriculture diminishes. More recently the proportion of the population engaged in energy intensive industries has declined whilst the proportion of those engaged in the processing of information has increased.

These socio-technological shifts of history have been marked by an increased intervention with the environment, the environment itself changing as a result. Initially the intervention was through limbs and other biological organs, later through tools and technology. The latter has changed in character, replacing many human (biological) modes of intervening with the physical environment by artificial means. In the last few years, several brain functions have been replaced and non-biological information processing is increasingly becoming important, as a mode of intervening with the environment, the environment itself being now increasingly defined as informational. Those who 'tend' this information environment, and the information economy were initially human but it is being done more and more by machines. The human component under the new industrial dispensation is tending to become marginalised and, to paraphrase Marx, human brains are becoming at least partial appendages of a machine dominated information environment.

The varieties of social arrangements we have described depend on the mode of intervening with nature and the social environment. They are a set of adaptations to the environment, some forms of adaptations to the environment, some forms of adaptation occurring early in human history and persisting till now. They are thus to be seen as the expression of a process of multi-linear social evolution in different physical and social environments of the world. The different social arrangements are in this sense, the result of an evolutionary radiation brought about by particular socio-technical arrangements as they adapt to particular social and physical environmental niches in different locations and in different times.

Once societies became sufficiently complex, stratification of these systems occurred at two levels. One level is at the major fault line of the social system, corresponding to class divisions based on the access to the means of production. The second is a subsequent division corresponding to the different professions and trades and based largely on technological criteria.

Persons occupying different roles in these social structures are socialised to operate on the cultural and information world that goes with the position. The stratification system provides the skeletal framework for the flows of information and culture.

There is a continuity of culture (common information and meanings) through time. Thus particular habits that may have been learnt from very early human existence, even from pre human days, are transmitted from generation to generation. Thus, some elements of the culture of three million years ago (say, on how to handle a chopper-type tool) existed in the culture of the neolithic revolution and parts of this culture continued in the later stages of human history. Thus, the world of information, knowledge and culture, has relatively speaking, some continuity in time. There is conservation of some information and culture down a transmission line extending through the various stages of human history, whilst new information is also created.

The common stock of information and knowledge shared by today's human society is larger than that of the 17th century, much larger than that of 3000 B.C. and very much larger than that of 3 million years ago. In addition, the different social strata, and professions to handle this increasing stock of information has grown continuously by differentiation of the strata, given rise to differentiation of the cultural lines. Down the flow line there is a continuity and conservation of some old information as well as the creation of the new, described for example by the evolutionary theorists of formal knowledge (Shapin 1982, Campbell et al.)

The historical trees of information and cultural transmission lines are very much like the tree of genetic evolution and are demarcated by the historically given social stratification systems that have grown over three million years of human or human-like history. Taking these social structures as a skeleton, socially acquired culture flows down the lines. And the flow is dynamic, responding to changes in the environment. The flow-stream connects the beginnings of human society to the present. It has conserved some information and culture from the early periods whilst continuously creating new cultural material.

I have described two flow lines of biological information. The first was the genetic, and when its limits of adaptation was being reached, a more flexible cultural flow line passing through a neural system was added. This tendency towards 'externalisation' of flow lines does not stop at the biological barrier. Just as the neural-cultural flow line developed 'externally', associated with the genetic flow line, other information flow lines can also develop as an external non-biological line associated with the biological.

The rudimentary beginnings of these exosomatic information lines can be traced back even to the earliest animals. But it developed and expanded only later with the primates and most elaborately only in association with humans. It is, as it were, only when the adaptive limits of the neural-cultural information line begins to be reached does it, in at least some functions 'spill over' into the non-biological, in a similar way, the neural-cultural line developed after 'spilling over' from the genetic.

'Evolution' of non-biological information

As man's brain enlarged and his manipulation of the environment increased, external adjuncts to his brain's memory as well as to his brain's information processing gradually grew. This was a new envelope of information physically external to the brain, in the same manner that the brain envelope was external to the genetic. At an early stage - probably even during paleolithic times - men would have used notches cut on barks, sticks or fingers of the hand as memory markers or as rudimentary information processing devices.

The invention of writing and recording of thought in symbols resulted in the storage, manipulation and transmission of a large amount of formal knowledge, than was possible in the earlier pre-urban stage. The accumulation of formal knowledge gave rise to storage of information (a 'memory' external to the human one) on papyrus, palm leaves and leather. Libraries now grew up in several cities. The information storage in words was initially done manually. But with the discovery of printing, the external memory of mankind increased quite dramatically.

Sometimes mechanical devices such as an abacus, pieces of sticks and other calculating aids, including fingers of the hand are used to partially process some of the information in these external memories. However, it is with mechanical electrical and electronic devices developed initially over the last 200 years and increasingly within the last 20 years that information processing outside the purely biological domain has increased dramatically. This has been accompanied also by an enormous increase in total memory storage outside both the human system and those of books.

Till about a decade or so ago the computer had been used for routine information processing in situations where the speed of computation helped surpass human abilities. A recent shift in emphasis is towards a more qualitative change, where several human abilities would be transferred into computer devices using Artificial Intelligence (AI) approaches, the aim of the Fifth Generation Computer programme.

In the early days of computers, in the 1950's, extravagant claims were being made both at a popular and at an academic level about the intellectual abilities of computers of the time. These were proved to be unfounded. Yet, surveying the computer scene of the last decade or so, one is struck immediately by the fact that considerable advances are now being made. There has been a growth of information processing devices having different degrees of intelligent characteristics including increasing abilities to learn, to recognise patterns, to mimic human professional expertise and so on. These have been done through developments in machine learning, expert systems, changes in machine architecture and new dramatic developments in hardware, specially the emerging 'connectionist' revolution. (Pask & Curran 1985, Feigenbaum & McCorduck 1983, Alexander & Burnett 1983, Isard 1986, Michi & Johnston 1985).

If the search for artificial intelligence can be crudely categorised by the dominant themes of the different decades, it would read as follows. The 1950's was the decade of the neural nets, the 1960's that of heuristic search, the 1970's of expert systems and 1980's of machine learning (Forsyth & Naylor 1986 p.1) and the 1990's probably of connectionist devices.

Space does not permit detailed discussion of the entire AI field but a very promising recent approach-the 'connectionist' one will be sketched because it brings an entirely new concept into computing itself. The connectionists, whose roots go back to a few decades ago, are of the belief that systems can be developed, which if connected together and endowed with sufficient complexity automatically process information. Connectionists are strongly influenced by examples from biological systems.

Connectionist approaches are still in their relative infancy and recent new developments in hardware suggest that connectionist devices approaching or even surpassing capacities of the brain may be technically feasible. These would be based on optical neural computers working on lasers and holograms. (Abu-Mostafa & Psaltis 1987).

Using volume holograms, that is, holograms in three dimensions, enormous interconnections are possible in these devices. A cubic centimetre hologram would allow for a trillion (10^{18}) connections alone by inter-connecting a million optical elements (ibid p.72). The human brain has roughly 1000 billion (10^{12}) nerve cells. But as each nerve cell is connected with as many as 10,000 (10^4) others, there are roughly 10 million billion (10^{16}) connections in the brain (Economist June 15th 1985 p.84). It appears that neural connectionist architecture using optical technology would increasingly dominate the future, and has the potential of reaching, if not surpassing biological devices.

These and other recent advances although significant, should however be considered only the beginnings of a long march (Rothfeder 1986). Once machines are created that exhibit intelligence, they can be ultimately designed to understand various aspects of intelligence and then design other systems that can be more intelligent. Artificial intelligence is an intelligence amplifier.

Exosomatic Intelligence in Perspective

An exosomatic information flow line that stores and processes information has grown quantitatively and qualitatively in recent years. Current computers have already surpassed the information handling capacity of elementary organisms. By the early 70's there were machines having an information handling capacity equivalent to the genetic information capacity of a single cell paramecium (Sagan 1978).

Increasingly, intelligent functions have been acquired by this exosomatic flow line, whilst it has become more extensive, more pervasive, and more integrated. These developments is best summarised in the words of De Solla Price (1983), 'We are being released from the limitations of the human body, one by one, and this new round may well be the final and ultimate liberation from the tyranny of being made from organic molecules, capable of evolution and perpetuation by the genetic code'.

Developments in the computer field are not to be seen as just only a tool, or a prosthetic device for any of our senses. The exosomatic flow line we have now acquired, has several autonomous aspects. The non-biological flow line is increasingly hiving itself off from the very direct links it has had with the neural (human) line and is consequently developing an increasing autonomy. Even 30 years ago, at the time when most computers were mostly used as sophisticated addition and subtraction machines, which only ran faster than humans, it was not possible for a human to check independently all the calculations in detail, simply because of the huge manpower and time constraints.

These constraints have of course increased considerably in recent years. The numerous computations now done by hardware (even in such-taken-for-granted devices as electronic watches, calculators, electronic typewriters, automatic cameras, automatic appliances and personal computers) would probably require the total human population harnessed in a tight organisational framework to give the same output at the same time.

The exosomatic information system is however 'still reacting explosively with its environment, going through a generation of change every few years, gaining a magnitude of power and cutting its costs in half every few years' (De Solla Price 1983 p.5).

The new information devices including those with intelligence, we should note, lend themselves to mass production. In short, mass produced intelligent devices would increasingly populate the world. They already do so in small products such as watches and appliances. This process would ultimately lead to a 'Xeroxing' process for machine intelligence (Smith and Debenham 1983). The artificial intelligence revolution consequently will (at least according to one scenario) 'upgrade and mass produce the finest intellectual decision making of society's scholars for public consumption' (ibid p.45).

With learning devices engrained in AI computers, and where the output of one AI machine feeds as input to another, without going through human intermediaries, humans will not be able even in principle, to be aware of the details of the modes of information processing (the reasoning, the logic, the sequence of steps etc.) that was involved in the resulting flow of information. Such flows would be 'opaque to human attempts to follow' the operation of the machine (Michel 1980). Increasingly the information line from the computer would take an autonomy of its own, specially with the extensive need for, and growth of new information sources and large data bases.

The growth of this formal knowledge outside the biological is to be seen in conjunction with the exponential growth of formal knowledge itself in recent centuries. Thus De Solla Price (1953) had shown a few decades ago that the number of scientific books and scientific journals grew at an exponential rate since the mid 17th century. This rate is higher than the rate of growth of population. This means that the flow of information outside the biological system is bound to grow at a higher rate than that within the biological as exosomatic devices increasingly take the load off human scientific workers (Open University 1971).

The exosomatic information flow line will thus increase its autonomy its interactions with the environment. This would lead to an increasing separation of its functions from the neural flow line, in the same manner that the neural flow line itself became separated from the genetic flow line.

Conservation and Novelty

The three flow line domains, the genetic, neural/culture and the exosomatic have certain common characteristics. It is useful to document a few of these.

All flow lines have a continuity from the past with a partial carry forward of information and modes of processing acquired from its early history. There is an innate conservativeness in the flow lines.

This is seen in the genetic line where past evolutionary successes are continued in the present (Eigen et al. 1982, Ayala 1978). This is also seen in the continuation of brain structures which once established, tends to persist (Walker 179-187, Jerison 1986). Also learnt, culturally transmitted information continues sometimes for thousands or millions of years; for example, that associated with making paleolithic tools.

Knowledge engrained in books etc. tend also to be preserved conservatively, as do other information in the exosomatic line in general. In the exosomatic line, new information is added largely on the basis of what has been stored and operated on before. Partly this line is built on earlier human generated information, but with machine lines developing an autonomy of its own, conservative tendencies continue in the exosomatic line.

The information flow lines not only has a past history and conservativeness built into it, but they also have a learning and creative aspect. This creative aspect has been called autopoiesis by Maturana and Varela (1975 & 1986), which is the process of continuous renewal of living systems, whilst maintaining the integrity of their structures.

This is seen in biological structures (Ho, Saunders & Fox 1986) in neural structures (Jantsch 1980 p.249), in the cultural learning patterns (Zelleny 1985), as well as in Artificial Intelligence devices (Zelleny 1977, Simon 1972). The creation of the new occurs in all three domains.

Speciation and Differentiation

Flow lines, as they proceed through time, differentiate and sub divide in the three domains. This is well documented in the literature on evolution in the genetic line including those on sudden disjunctures (Gould 1982).

Differentiation of information flow lines also occur in the human realm as changes occur from the earliest human social organisation, the hunting band, through the tribe to the more recent manifestations of classes corresponding to different stages of human evolution. Smooth gradual differentiation (speciation) of human sub-cultures is interrupted by sudden jumps which occur by the introduction of new tools, new forces of production of new modes of interacting with the environment.

Such disjunctures occur in the transition from paleolithic to the 'neolithic', from the Neanderthal man to Cromagnon (Torrence 1984, Dannel 1984), from the neolithic to copper, to bronze to iron, and later to the emergence of other new technologies such as that associated with industrial capitalism. Each such disjuncture is accompanied by a major re-ordering of the macro-social structure, with new classes and new social groups; which are considered under the heading 'social stratification' in conventional social science treatments. These new classes and social groups are new sub information flow lines replacing earlier ones, or are a re-ordering of the earlier flow lines.

Speciation and differentiation could result in further sub divisions within a macro flow line. Thus within a particular class structure for example, flow lines corresponding to micro level occupational sub cultures could emerge. Such sub cultures would correspond to the social divisions brought about by professions. These exist within the major fault lines of society, (the class structure) associated with a given mode of production.

Speciation also occurs in the case of information held in non-biological storage systems. Even in the early forms of storage, there are copying errors from manuscript to manuscript resulting in a wide variety of interpretations of the information. Such copying errors could conceivably lead sometimes to new sub lines, as humans working on such error-filled material giving rise to new interpretations and new knowledge linkages.

The growth of the exosomatic flow lines would initially follow the broad contours laid down by human information systems. Thus, there would be sub-lines of information storage and processing following disciplinary boundaries and library classification systems that have been laid down by humans. These same disciplinary boundaries exist in such AI devices as expert systems, because they follow by and large, the same disciplinary approaches.

With continuous exchange of information between different exosomatic flow lines and specially with the intelligent processing of information filtering this exchange, new fault lines and new flow lines in the information flow would emerge. This is a necessary future outcome in the extensive computer and communication interactions that we have already documented. Communication between exosomatic flow line do not occur at random, just as genetic flow lines do not interact in random.

Apart from speciation there are also processes of integration in the three domains brought about by information held commonly such as in a gene pool or in a social community. There are also cross flows of information such as between flow lines in the social system that help tie in the social flow lines together.

Rapid Information Growth and Discontinuities

With the emergence of a particular niche in an environment it appears that the flow line changes very rapidly to fit into, as it were, a vacuum. Thereafter a plateau is reached, very little change occurring thereafter. This seems to be reflected in the new evolutionary model of punctuated equilibrium (Eldredge & Gould 1972,1977).

A similar process seems to occur in the evolution of brains where once an ecological niche is available the neural system expands rapidly, later settling down to a constant form and size (Jerison 1986). Similar processes are seen in social and cultural flow lines. As the social and cultural flow lines associated with human societies move along the path of history, they reach sudden disjunctures. These disjunctures arise from new relations to the social environment or to the physical environment.

A new mode of intervening with nature results in new patterns of social structure, of cultural flow lines and new socio-cultural dynamics. These are seen in the changes from say the paleolithic, to the mesolithic, to the neolithic, to the incipient agricultural, to advanced agricultural 'feudal' systems, to industrial capitalist and so on.

Cores of Information and Response Times

In the three flow lines one should note that responses and adaptation to the environment becomes, or has the potentiality to become, quicker as one proceeds from genetic to neural to exosomatic; that is up as it were, a phylogenetic sequence.

The different flow lines cores do not exist unconnected and unattached to each other. They are inexorably linked. Thus, the process of externalisation of the information envelopes is such that given sufficient evolutionary time, and suitable environmental pressure the externalising of the genetic to the neural flow line and from there, to the non-biological must occur. It is a continuing process of reacting with, and adapting to the environment resulting in more flexible and quicker responses to the environment. The tendency towards further elaboration and evolution of information is therefore but a necessary outcome of an almost inexorable phylogenetic process.

The length of information lines that remain unchanged are different in the three domains. Changes in the genetic one takes geological time, in the cultural one a shorter time and in the exosomatic information it is virtually in 'real time'. The length of information carried over unchanged thus becomes shorter and shorter in the successive flow envelopes, from the genetic to the neural to the exosomatic.

Cognitive Structures

Every information line has a window to the world outside, which in a sense is its own 'subjectivity'. This 'subjectivity' has been discussed in biological systems by Plotkin (1982) and others. Here the living process is seen as one, where the 'world' is being continuously structured (Oyama 1985). Biological organisms in that sense have different 'world views' through their genetic systems (Greene 1985, Morin 1981).

Similarly different world views are also constructed in different brain systems (Jerison 1986). Different animals thus perceive the world differently through their hardware (ibid) In the cultural realm too, this aspect of different subjectivities of the flow line has been amply documented by writers on the social construction of reality, on the social construction of science as well as by those writers on class consciousness such as Lukacs.

This same phenomenon appears in the exosomatic line, where the world is sampled in a particular manner through a particular spectrum of sensors and particular forms of processing of information. Thus every system of knowledge whether it be genetic, cultural or exosomatic operates by selecting meaningful data whilst rejecting others giving each flow line its own particular identity, its own 'egocentricity' (Morin 1985 p. 135).

Rigidities

The evolving flow lines of information have also certain rigidities. Earlier acquired information, the inner flow lines, sets through their world views the guidelines and limits, for the responses of the later and outer, more flexible flow lines. The inner genetic core thus sets limits to the variety of behaviours and environmental responses possible from the neural. The inside core is thus like the hand inside a glove puppet. The information of past epoches (say as encoded in genes in the innermost flow line) exists as past history, congealed as 'world views' that guide the more flexible later flow lines, including those associated with the exosomatic line.

If one were a 'passenger' through time accompanying the flow of information, one would have a ride through time something like a roller coaster. One can go down any phylogenetic tree, genetic, social or exosomatic, and 'see' the outside world unfolding and changing, becoming more varied, different apertures emerging as cognition and gestalts change. In the process, world views enlarge and occasionally contract. This could best be pictured like a ride down a roller coaster with a video screen in front. The screen itself contracts and enlarge. The screen's images change. Colour is added. Sound is added. Its acoustic spectrum changes, smells are added and are changed. There are also turning points in the roller coaster which result in different pictures on the screen, new cognitive maps are projected as new species come into being. The roller coaster goes up and down, turns and bifurcates corresponding to environmental changes.

What I have briefly summarised in this section are only some of the general common characteristics of the flow lineages of information and of complexity. These common characteristics include conservation and creativity, gradual change and disjunctures, speciation and integration, co-evolution, rigidities and flexibilities, different response times and relative lengths of flow lines and egocentricity of world views in flow lines. These facets have also been dealt with, using sometimes different terminology, in the many formal disciplines that cover the 3 to 4 thousand million years of history of the different flow lines on earth.

Deep Dynamics of Flow Lines

The dynamics of flow lines are operated upon at a deeper level by more fundamental laws. These are the laws of entropy, specially the thermodynamics of open systems as described by Prigogine (Prigogine & Stengers 1984) and others. The various segments of flow lines that we have described have been analysed from the Prigogine perspective using the thermodynamics of open systems by several writers. These studies include those describing the biological realm (Brooks and Wiley 1987), the central nervous system (Bauman 1975, Freeman 1975), socio-cultural phenomena like urban dynamics (Allen 1985), evolutionary stages of society (Taylor 1976, Malaska 1985) and exosomatic devices (Simon 1962).

Entropic forces and the thermodynamics of open systems operate on the flow lines. Although conceivably open systems characteristics can be discerned in simpler objects, information flow lines have the complexity to make open system transformations a major feature. Only in the flow lines do significant emergent phenomena develop and dissipative structures operate.

Implications

The above brief and highly simplified sketch of the evolution of information ties in the two information 'bundles' that we are interested in this meeting, namely the artefactual and the social, the computer and the human.

It is not my aim here to discuss in detail how this perspective can be brought in to discuss the detailed relationship between the two, specially in the design of human centred artificial systems. I should note here that, the relationship of the theoretical perspective I have sketched to the design of human centred systems is in many ways analogous to the relationship of the theory of biological evolution to the individual design of bio-technological systems. Evolutionary theory could provide a general background knowledge, but not necessarily a detailed knowledge of a 'how-to' kind. Yet, I wish to make an entry to the practical problem of human centred systems by posing a set of questions.

1. What has been described here is the long term diachronic flow of information. Yet, social and artefactual devices function and survive on a day-to-day individual basis, (as opposed to the survival of a lineage, a species) on the basis of the interactions of the 'device's synchronous information. It is the latter, that provides the detailed day to day interaction with the environment, and the dense feed back loops (both internal to the system and external to it) that help maintain its identity and exist in stable relationship with its surroundings.

This synchronous information has been described in the classical literature of cybernetics and information systems generally. The question to be posed is, - what are the relationships of this type of synchronous information links to the diachronic variety and how do they effect each other?

How does also synchronous information flows of one lineage interact with that of another, outer lineage. That is, for example, how does cultural information interact with those of artefacts? How does one describe this and how can one intervene in these interactions.

2. Artefacts have also been described recently as evolving, as technologies change to fit into market and other environmental niches. A considerable literature has grown up that describes the fine tuning of products and technologies as they adapt to a variety of niches. How can this evolution be described in terms of the general evolutionary scheme that has been described here? Once such a description has been made, can one then go on from there to describe how interventions can be made in new devices so that a greater environmental fit could be arrived at, that is a greater fit between one flowline with another, the human with the artefactual?

3. The classic literature of socio-technical systems described how the social system was interconnected with the technical-and vice versa. Can this literature now be reformulated in the terminology of the interaction of two sets of information lineages, to yield a fresh insight?

Thus, could the early work of the Tavistock Institute or of Joan Woodward be reformulated in this terminology? How would such categories as process industries or batch production, specially when it relates to the newer information technologies be then seen?

Can ergonomic problems be reformulated using this terminology? How about job enlargement, design of jobs, factors of motivation? Could an entry point for the human factor be through psychology, specifically social psychology?

4. The manner in which social perception is historically constructed through essentially social-evolutionary mechanisms has been described by writers such as Vygotsky and Mead who are now having a strong revival. In such a scheme 'culture' is the socially transmitted information lineage. How can industrial culture be then viewed from such a perspective? How could it be reformulated to give a fresh insight from a long range evolutionary perspective?

5. The different information lineages give different 'views' of the world, selective windows of perception of the world outside. This is so whether the perception is through artefacts or through the social/cultural lineage. Would a study of these different 'egocentricities' in perception provide a useful entry point to design human centred systems?

6. What in fact is human centred design of artefacts in the terminology of the evolution of information? A perspective from one flow line to another? What are the long term evolutionary implications of such a perspective? Would the attempt to match the two through better ergonomics, through better participation in design of the humans involved in their use, lead in the long evolutionary run to a better survival of the newer artefactual lineage? A better match of the human to the faster evolving artefactual lineage, leading ultimately, in evolutionary terms, to the latter's long term differential growth and survival?

References

Alexander, I & Piers, B (1983) *Re-inventing Man* Haxmondsworth, Penguin Books

Ayala, F J (1978) The Mechanisms of Evolution. In *Scientific American*, September 1978

Baumann, G 1975) Das Kunstliche Nervensignal Synthese einer Zellfunction. In *Neue Zurcher Zeitung*, 29 January 1975.

Brooks, D R & Wiley, E O (1986) *Evolution as Entropy: Toward a Unified theory of Biology*, Chicago & London, The University of Chicago Press.

Campbell, D T et al (1986) Evolutionary Epistemology Bibliography. In Callebaut, W & Pinxten, R (eds) *Evolutionary Epistemology: A Multiparadigm Program*, Boston & Dordrecht, Reidel.

Dennel, R (1986) Needles & Spear Throwers. In *Natural History*, October 1986.

De Solla, P D J (1963) *Little Science, Big Science*, McMillan, London.

De Solla, P D J (1983) End of the Naked Brain. In *Interdisciplinary Science Reviews*, Vol.8. No 1, 1983.

Dobzhansky, T (1962) *Mankind evolving: the evolution of the human species*, New Haven, Conn. Yale University Press

Eigen, et al (1982) The Origin of Genetic Information. In Smith, J M (ed) *Evolution Now : A Century after Darwin* Nature, MacMillan , London.

Eldredge, N & Gould, S J (1972) Punctuated equilibria: an alternative to phyletic gradualism. In Schoff, T J M (ed.) *Models of Paleobiology* San Francisco. W.H.Freeman, pp.82.115.

Feigenbaum, E & McCorduck, P (1983) Land of the Rising Fifth Generation. In *High Technology*, June 1983.

Forsyth, R & Naylor, C (1986) *The Hitch-Hiker's Guide to Artificial Intelligence*, New York, Chapman & Hall/Methuen.

Freeman, W J (1975) *Mass Action in the Nervous System: Examination of the Neurophysiological Basis of Adaptive Behaviour through the E.E.G.*, New York, Academic Press.

Gould, S J (1977) Darwin's Delay. In *Ever Since Darwin: Reflections in Natural History*, New York and London, W.W.Norton pp21-27.

Ho, M-W , Saunders, P & Fox, S 1986) A new paradigm for evolution. In *New Scientist*, 27 February, 1986.

Isard, S D (1982) Synthesis of Rythmic Structure. In *Proceedings of the Institute of Acoustics conference*, Bournemouth, ppE.1.-E.4.4.

Jantsch, E (1980) *The Self-Organising Universe*, Oxford, Pergamon Press.

Jerison, H J (1985) Paleoneurology and the Evolution of Mind. In *Scientific American*

Lumsden, C J W & Edward, O (1981) *Genes, Mind and Culture: The Co-evolutionary Process*, Harvard University Press, Cambridge, Massachusetts & London, England, xii.

Malaska, P (1985) Outline of a policy for the future. In *The Science and Praxis of Complexity*, United Nations University.

Maturana, H R & Varela, F (1975) *Autopoietic Systems*, Report BCL 9.4. Urbana, I11; Biological Computer Laboratory, University of Illinois.

Maturana, H R & Varela, F (1986) *The Tree of Knowledge: The Biological roots of Human Understanding*, Shambhala.

Michie, D & Johnson, R (1985) *The Creative Computer* Pelican, London.

Morin, E (1981) Self and Autos. In Zeleny, M (ed) *Autopoiesis - A Theory of Living Organisation* New York, North Holland.

Morin, E (1985) Blind Intelligence. In Shlomo Giora Shoham & Francis Rosensteil (ed) *And He Loved Big Brother*, Lourdes MacMillan, Council of Europe.

Oyama, Susan (1985) *The Ontogeny of Information*, Cambridge, Cambridge University Press.

Pask, G & Curran, S (1982) *MICROMAN: Living and Growing with Computers*, London, Century Publishing Co.

Plotkin, H (ed) (1982) *Learning, Development & Culture*, Chichester, Wiley.

Rose, S (1976) *The Conscious Brain*, Penguin, Harmondsworth.

Rothfeder, J (1986) *Minds over Matter*, Sussex, Harvester Press.

Sagan, C (1978) *The Dragons of Eden*, New York, Ballantine Books.

Shapin, S (1982) History of Science and its Sociological Reconstructions. In *History of Science*, Vol.20 (1982) 157-211.

Simon, H (1962) The Architecture of Complexity. In *Proceedings of the American Philosophical Society*, 106, 6.Dec .1962.

Smith, G W & Debenham, J D (1983) Mass Producing Intelligence for a Rational world. In *Futures*, February, 1983.

Taylor, A M (1976) Process and Structure in Sociocultural Systems. In Erich Jantsch & Conrad H.Waddington (ed) *Evolution and Consciousness: Human Systems in Transition*, Massachusetts, Addison-Wesley Publishing Co.

Walker, S (1985) *Animal Thought* Routledge & Kegan Paul, London.

Zelleny, M (1979) Cybernetics and General Systems - A Unitary Science. In *Kybernetes*, 1979, Vol.8, pp.17-23, Thales Publications (W.O.) Ltd., Great Britain.

Zelleny, M (1985) Spontaneous Social Orders. In *The Science and Praxis of Complexity*, United Nations University.

A Conceptual Framework for the Concept of the Artificial

Massimo Negrotti

The Role of Isomorphisms in the Development of Science and Technology

The systems approach to natural and social sciences has often emphasized the existence of isomorphic aspects of reality in order to support the bases of a general systems theory.

The concept of isomorphism, beyond its proper original areas of mathematics and chemistry, does in fact offer the possibility to understand the mutual similarities of things, allowing interesting taxonomies or analogical descriptions and explanations in many fields, ranging from biology to sociology.

Instead of concentrating our attention on the differences among objects, the search for isomorphisms emphasizes the key role of similarities in our understanding of the world.

One of the most immediate ways to find out isomorphisms is that of observing directly the reality by means of a sort of pattern matching strategy. A conceptual frame for isomorphisms can be based upon the following propositions:

1. Observable isomorphisms depend on the level of description and of observation; presumably this applies also to isomorphisms that aren't observable but only conceptualised.

2. To recognize isomorphisms, humans should be able to capture the right amount of information in a right and limited 'tuning' range often drawing it from a random set of elements.

3. The variable amount of intrinsic or objective reality of isomorphisms suggests the idea of a limited range of possibilities available to nature and to man in creating forms, architectures and solutions to problems of adaptability. In other words, natural dimensions constrain and partially model the possible solutions. As far as man is concerned, the range of possibilities seems to be determined also by the fact that what we observe and what we think of are inter-dependent phenomena which are linked together. In order to achieve a higher degree of mental freedom from the constraints of this loop, human culture has developed very slowly the necessary intellectual tools, such as logic, mathematics and abstract theory.

4. From an evolutionary point of view, isomorphisms appear to have developed according to three models:

IN *Intra-Natural:* natural systems develop *as if* they were drawing their forms, architectures or functioning features from other natural systems.

NC *Natural-cultural:* both intentionally and objectively, cultural artifacts assume forms, architectures or functioning features drawing them from natural examples or having to submit to the same natural constraints.

IC *Intra cultural:* both intentionally and objectively, cultural artifacts assume forms, architectures or functioning features drawing them from other cultural examples or having to submit to the same cultural constraints.

Of course, the isomorphisms of the IC kind are today much more widespread than the NC ones, owing to the intense *technological environment* we have built around us. In this way, the development of this intra-cultural process has led to the rise of objects or systems which no longer show similarities with nature, opening the door to the discovery of a true *artificial reality*.

It is even presumable that this situation should produce a fourth kind of isomorphism which could arise from the need to adapt natural systems to artificial ones. Anyway, this is surely the case with humans who survive not only in the natural environment but also in the technological one.

Artificial Intelligence (AI) as a Backward Oriented Search for Isomorphisms

As far as an intentional way of working (intentional adaptation) was concerned, the strategy of isomorphism was successful because it aimed to ensure the best and most direct adaptation to the environment. Nevertheless it works well only in the first steps of making technology, owing to the fact that we can reproduce only those aspects of reality which we can see or understand by means of simple models, and if we have to achieve simple targets (for instance that of protecting ourselves from the rain, that of penetrating a fluid with the lowest waste of energy, or that of building up an effective communication or transport network).

Furthermore, beyond certain limits, the reproduction of given systems would imply not only the reproduction of static features but also the reproduction of underlying dynamic processes, and this is almost impossible or even useless.

The roots of the artificial reside just in the attempt to capture some features of real systems, if possible. In this process, the similarity with the real system soon becomes of little importance in se and remains strategic only if and so far as it allows an *effective achievement of higher performance*.

An artificial kidney does not necessarily replicate the real dialysis which happens in the real organism, and, in the same manner, a computer does not necessarily perform math calculations in the same manner as the human mind.

Often, the artificial assumes the natural system as a starting point but in a short time it becomes something autonomous, ridding itself of pure reproduction and achieving features very different from the exemplar, sometimes enhancing them, sometimes reducing their power.

From an historical or evolutionary point of view, the most effective technology emerges from the abandonment of the isomorphic targets, i.e. when it becomes a discipline in itself, open to intellectual landscapes which have no reference in the natural world. If this is a fact, obviously the problem of the cultural adaptation of man to these new landscapes becomes central as a true evolutionary issue. It is still more crucial if the density of the technological, or artificial, environment grows in such a measure that it becomes the principal source of intentional or objective isomorphisms.

From the above statements it follows that AI is a kind of research that, if pursued with the aim of reproducing the human mind and natural ways of reasoning, goes back to a pre-modern technological culture, towards a naive search for isomorphisms. AI appears from this point of view, to be a search based on the persuasion that it is possible and convenient to find out similarities between a very complex natural system (the mind or the brain) and a project for an cultural artifact (the computer). But, in doing so, AI encounters serious additional limits since it has to deal with almost non-observable phenomena and, therefore, its claim of reproducing forms, architectures or functioning features of the real system remains a pure declaration.

The only persuading isomorphism we can find is based on the fact that both brains and computers process information, but it would be too weak a starting point if we were to describe and explain, on this basis, the whole set of knowledge performances of the human mind and build up a human-like machine assuming this very limited and partial model. Many artificial systems (like submarines, air crafts or radars) share with natural exemplars some basic features, but their development and their enhancements have soon become very different from the initial isomorphisms assumed as starting points, revealing this way their true artificial reality as a new field of research. This is true both for the so called *symbolic-AI* and for the *connectionist or neural nets-AI* (on the wide debate about the most suitable models for reproducing the mind's performance and on its theoretical implications, see Nelson, 1989).

In fact, in the case of symbolic-AI, the computer programs are, in the end, isomorphic only with the cognitive models built up just for giving the machine their kind of intelligence, and not with unknown and non-observable real systems and processes.

In the case of neural nets the discovery that large communities of simple elements are often able, in natural environments like a community of ants or a neurons system, to exhibit intelligent collective behaviour, is not enough to support the claim that the achievement of a particular behaviour from a net demonstrates the reproduction of the basic features of the brain.

Though in the neural net example the isomorphism with reality is much more based on observable real phenomena, it is clear that this isomorphism captures only the surface of very complex systems. It works well as a starting point and even as a solution for some kind of problems, but whatever enhancement we may expect from this tradition of research it will derive from the abandonment of the isomorphism strategy.

The community of robots that Maes and Brooks at MIT are developing for working on the moon in the next decades, is a good example of an isomorphic starting point, since it assumes the behaviour of a population of ants as a model, but it is very plausible that further enhancement in performance of the robot community will be achieved without any need of similarities with the biological model.

The above considerations testify that AI is, at the same time, a discipline which deals with very complex and highly developed tools and a *technology* oriented toward a very backward philosophy of working. This is partially due to the fact that we are only at the beginning of a new tradition of research and, therefore, the strategy of the isomorphisms (apart from their well founded nature) acts now as the only possible mode of work as happened in the past for the first steps of every technology. But it is due also to the insistence on the view that biological intelligence is the only one worthy enough to be reproduced, neglecting, in this way, the actual intelligence of the machine as such, that is as an artificial reality.

For instance, the debate on the *real* nature of what happens in a thermostat is only intended to establish whether it has 'opinions' or not, as exemplified by the famous and amusing discussion between Searle and MacCarthy, but from such a way of discussing we will get no additional knowledge on what an opinion is, or on what kind of intelligence is embedded in a computer program which controls a large set of sensors. After all we should understand that we are giving the machine more and more power in organizing and controlling our life and, in so doing, it isn't important whether the machine has or has not human-like opinions, but it is of strategic importance to understand its logic or intelligence.

Conclusions

The concrete advancements of AI or computer science are amplifying our artificial environment, despite their lack of similarity with the human mind or brain. Perhaps we are entering on a new phase in which the machines will become more and more intelligent just by means of abandoning the model of isomorphism with human intelligence.

This implies that we should be aware of what the machine's intelligence means and what it involves in terms of cultural adaptation and evolution, regardless of its similarity to the human features. Human-machine systems are not to be conceived as a homogeneous whole but in their irreducible heterenity. Only starting from this premise which emphasizes the variety of the kinds of intelligence can we get from the phase a true cultural enhancement.

Otherwise, the illusion of the isomorphism could suddenly explode bringing us back to the same critical situation in which we now find ourselves with regard to the natural environment, which we have treated with too friendly a philosophy, really ignoring or neglecting its actual nature.

References

Emery, F E (ed.) (1972) *Systems Thinking*, Penguin Books, London.

Miller, J G (1965) *Living systems: Cross-Level Hypotheses*, Behavioural Sciences 10, 3-4

Negrotti, M (ed.) (1990) *Understanding the artificial*, Springer-Verlag, London, in press.

Nelson, R J (1989) Philosophical issues in Edelman's neural Darwinism. In *Journal of experimental and theoretical artificial intelligence*, 1, part 3, pp.195-208.

Dialogue and Design of Computer-Based Technology

Satinder P. Gill

The paper discusses the role and significance of tacit knowledge in dialogue for the design process of computer based applications. In particular knowledge based systems and multi-media communications. In both these technologies the nature of the communication amongst designers, and designers and users affects the nature of the end product and its use. In the case of knowledge based systems, the focus on dialogue is on the knowledge engineering process. In the case of multi-media communications, the focus is on the dialogue amongst designers in a design team who are at various times, both designers and users in relation to each other.

The paper will focus largely on the case of knowledge engineering. This involves a communication between a *designer* and a *knowledge source* (e.g. a doctor, a lawyer). Within this domain I pose the questions: 1) what are the assumptions made about *what is knowledge* by those who undertake knowledge engineering?; and 2) what implications does this have upon the effective design of a system, i.e. what effect does this ave on human-machine communication - how it affects the user. Both these questions are concerned with the nature of tacit knowledge - that knowledge which is known, understood, communicated, yet not formalised in procedural or propositional form.

Over the past few years in the field of knowledge based systems design in Britain, the focus may be described as having been being machine-centred, leading to a Tayloristic design process of 'knowledge acquisition', 'knowledge representation' and 'man-machine interaction'. Metaphors such as 'acquisition', 'engineering', and 'elicitation' make the assumption that the nature of skill can be reduced to structured 'data', 'information' and processes making it 'operational'. The knowledge acquisition process, in this case, is seen as 'knowledge engineers' explicating the implicit knowledge of domain experts and developing general models of expertise, not of understanding the nature of their practice. No account can be taken of human context i.e. personal experiences, culture, values, ethics, aesthetics, etc.

To do the latter, it is proposed, that an understanding of experts' tacit knowledge is fundamental, as it is this which makes the public knowledge, and shared knowledge meaningful to each member of the working life environment. The focus has to return to the purpose of design. This is defined by the purpose and use of the knowledge based system, be it by doctors, engineers, nurses, etc. It is not a matter of considering the 'cognitive abilities' of an expert as an abstraction of the human mind, but a matter of understanding the practice of expertise of a person in the context of the working life environment.

To tackle these concerns in my research, I needed methods of analysing dialogue which would demonstrate the role and the significance of the tacit dimension in communication. I chose to undertake discourse analysis of both visual and verbal communication in a design process. I am using this method in my current case study of the design team of a multi-media communications infrastructure in the research centre of a multi-national information technology company in the UK.

Discourse analysis, in the field of social psychology, enables the qualitative analysis of communication. I am using it to identify various kinds of knowledge in communication amongst the design team, with a particular focus on the tacit dimension of the discourse. This involves the analysis of, for example: Roles, whether these be in the group, personal, social or organisational; Styles of communication, e.g. manner of speaking; Topics, eg on issues such as 'privacy' and 'control'; communication breakdown - how it occurs and how, if so, it is mended. Such categories for analysis of communication involve the consideration of aspects such as power, self-expression, purpose, images, interpretation, and common language, etc. For example, 'common language' could be social/organisational, or of the group, where it is the publicly shared or agreed upon language use. These aspects of communication will be discussed in this paper. For example, the study of body language enables the identification of tension in dialogue or where there is ambiguity in the communication.

Although the technology of knowledge based systems is different in nature from that of multi-media communications, the fundamental proposal of the research is that the nature of dialogue in the design of technological systems has embedded in it the conceptual development of the system and affects its practical construction. Hence the significance of using discourse analysis to identify the dynamics of the kinds of knowledge and knowledge structures in the process of communication, with a particular concern to show how the communication of tacit dimension is transformed into 'action', where action is the 'decision-making' in the design process.

The discussion in this paper will follow through the ideas in experts systems design, the field of cognitive science and Artificial Intelligence (AI), and is based upon case studies from the Swedish Centre for Working life and the British Alvey programme (Gill, SP 1988), and two case studies undertaken with a multi-national computer company in Britain and a British firm of underwriters.

Expert Systems and Computational Psychology

The aim of designing expert systems has mainly been to simulate 'those elements of human reasoning in which human experts are highly competent' (Kidd 1987). There is now an interesting shift to viewing the human and the expert system as a 'joint cognitive system' where AI techniques may be used to support 'those aspects of reasoning at which most humans (experts or not) are weak, as a result of their in-built cognitive limitations' (Kidd on Wood 1986).

Computational psychologists 'see the mind's rich tapestry as a mass of psychological patterns woven from computational thread' (Boden 1989). They take a functionalist approach to the mind, defining mental states in terms of their causal effects on other mental states and behaviour, and seek to identify mental processes with specific effective procedures. The mind becomes a representational system, and psychology becomes the study of the various computational processes whereby mental representations are constructed, organised, interpreted, and transformed. This computational conception of mind have been changing with changes in hardware and software capabilities of the available technology. For example, since the 1960's Cognitive psychology has advocated that the mind is made up of memory buffer stores and registers, parallel processors in the 1980's, and currently, parallel distributed processing - connectionism. The belief is that causal information processes justify ascribing the properties of meaning, representation and symbolism to the machine. It is argued that to compute and to understand are fundamentally similar processes, therefore computational psychology can give a science of the mind. Boden (1989) does accept that if one cannot attribute understanding to computer programmes then one cannot use theoretical concepts to be drawn from an AI context to explain the understanding which is possessed by people. This does not, however, change the fact that computational psychology is the culture of mind models, and abstract structural symbolism. It is a paradigm that cannot account for human emotions, motivation, and beliefs, as these human qualities are not rational.

The paper argues that these human qualities qualities, together with the vast experience humans have in their lives, are an essential part of human dialogue and behaviour. It is these qualities and experiences that enables experts to deal with uncertainty for effective decision making in working life. On the other hand, *Knowledge* which is objectively defined in the vein of the *rationalistic-rules model* approach of AI, imposes a quantification of knowledge and skills in *knowledge engineering* (Gill 1988).

There is an argument (Diaper 1989) that 'expert systems only need to behave as if they contain the knowledge of human experts, and there is no requirement that they possess, in their internal architecture or processes, anything like the true psychological mechanisms of real people', as they are basically an applications-based technology. The suggestion is to stop the focus on emulating human thought processes, which is something we still have no true understanding of. Instead, the focus of research should be on emulating domain *experts' behaviour*. The justification for emulating behaviour is that this can be observed. However, the same process of 'extracting' and 'making explicit' human domain expert's knowledge, (knowledge elicitation), and then translating this into a form suitable for programming into a knowledge based system (knowledge encoding), is to be followed. What is more, cognitive psychology uses the behaviour of human subjects as a major source of its evidence for constructing models of the human mind, who are usually tested in laboratories where the stimuli and response constraints on subjects can be recorded and controlled. This laboratory experimentation poses great problems when relating the results of this method of analysis to experts in their work environments who are practising their skills. This involves relating to other people. People's working practice and therefore skills also, cannot be considered on an individualist basis. Expertise exists in its practice, and decision making involves dialogue, whether the domain is nursing, medicine, underwriting, consultancy, weather forecasting. And dialogue, has essentially arrational human qualities. There are two problems posed by psychology for the cognitive scientists: the relationship of consciousness to psychological processes and its representation in the form of human behaviour. However behaviour such as speech or drawing cannot represent all that we are going through in our consciousness. Hence cognitive psychology depends on inferences drawn from behaviour.

Knowledge Engineering and Knowledge Acquisition

The core concern of knowledge engineering and the whole design process of knowledge based systems is still on knowledge 'representation', be it of behaviour or mind. This is the constraint of the symbolic representational infrastructure of the computer machine. Propositional representation of reality is the objective. It becomes important to identify end users so that the knowledge elicitation process can be 'targeted' towards them through the use of various elicitation techniques. For example task analysis techniques eg Command Language Grammar (CLG), Task Action Grammars (TAG), Task Analysis for Knowledge Descriptions (TAKD), may be used to identify actual requirements of users, and the possible range of tasks and sub-tasks the expert system or KBS might be expected to perform. Needs of users are determined through applying specific techniques, rather than through dialogue.

In the debate on knowledge representation, knowledge is divided into being *declarative* or *procedural* (Winograd & Flores, 1986). Arguments range from 'natural language is largely declarative', to 'psychological representation is always procedural', although peoples' behaviour appears in declarative form.

The knowledge engineering community has long seen the inability in experts to express their knowledge and skills in procedures in natural language, as a human shortcoming, hence the emphasis placed by many of them upon techniques. One of the most problematic areas identified by knowledge engineers is the explication of tacit knowledge. Verbal declarative knowledge (what experts say) of the experts has to be formerly analysed so that it can be transformed into procedures using, for example, laboratory experiments and protocol analysis using card sorts and laddered grids. So the debate continues in both knowledge engineering and Human Computer Interaction (HCI) about the problems of verbal knowledge and human behaviour, with the focus still remaining on procedural representation derived from an analysis of declarative knowledge.

In short, the process of knowledge acquisition, as defined by Kidd (1987), and of which there is general consensus in the systems development community, involves the following:

1. Employing a technique to elicit data (usually verbal) from the expert.
2. Interpreting these verbal data (more or less skillfully) in order to infer what might be the expert's underlying knowledge and reasoning processes.
3. Using this interpretation to guide the construction of some model or language that describes (more or less accurately) the expert's knowledge and performance. Interpretation of further data is guided in turn by this evolving model.

In the universalistic/rationalist conception of knowledge, knowledge becomes equivalent to the concept of data, as is clearly illustrated by Kidd's description. It becomes an abstraction of reality without any account of the human context i.e. personal experiences, culture, values, ethics, aesthetics, etc. For example, directed by the cognitive-modeling approaches, natural language research 'ignores any social, cultural and personal contexts which are essential for communicating, gaining a language and its meaning' (Gill 1988). It is not surprising therefore that this also affects research on knowledge and skill acquisition. The problem becomes not of 'understanding the skills and expertise' of the 'expert', but of devising 'techniques to elicit data'. Hence the use of the metaphor 'engineering' becomes acceptable for describing and prescribing the process of acquiring knowledge. By engineering I am here referring to machine-centred approaches of the engineering world, as explained by Rosenbrock (1988). The machine-centred approach leads to the impoverishment of human abilities rather than empowering them.The knowledge engineer then becomes the expert required to interpret the data to infer underlying reasoning processes. There is no mention of involving the 'domain expert' in this interpretation process, yet this interpretation is going to be the basis of the 'model' or formation of a 'language' to describe the expert's knowledge and 'performance' (see Ehn 1988, Cooley 1988).

The limitations and constraints of the machine-centred knowledge engineering process above requires an alternative approach which will be able to make for a deeper understanding of skills, and thereby more effective purposeful design. It is proposed that this may only be achieved through a focus on the nature of tacit knowledge, practice, and human-human dialogue.

Tacit Knowledge

The concept, *tacit knowledge*, was coined by Polanyi in his work on 'The Tacit Dimension' (1966), and developed in 'Personal Knowledge', where in essence, he described it as that which we cannot express, but that which we know. In the world, there is formal, explicit knowledge which can be propositionaly defined. However, the propositional knowledge only has meaning when we know how to use it, make sense of it. This entails its practice. For example, a chair make sense because we sit in it. The practice of propositional knowledge makes for its understanding. This is tacit knowledge, that which gives propositional knowledge any meaning or purpose. This extends from the objects in our world, such as a chair, right up to our skills and our language. In terms of experts in the workplace, tacit knowledge can be described as entailing:

- personal knowledge: that which we gain from our personal life experiences eg family culture, school, friends, i.e social values, beliefs etc.;
- experiential knowledge: that which is specific context based, in the workplace, eg work colleagues, group culture, organisational culture etc.

When a novice enters a work place, there are rules to be learnt and followed, tasks to do, skills to be trained in. It is the personal knowledge and, say, knowledge gained from previous work experiences, which the novice will use to interpret, understand, adapt, and gain an understanding of the world of work. Through this apprenticeship process the novice will become an expert (Dreyfus 1986, Cooley 1987) who will have the ability to adapt the rule, and gain confidence to break rules. This is the creative art of any expert. This tacit knowledge is so much a part of our identity that we may not be conscious of it in the explicit sense of the meaning of 'to know'. That is, when a nurse cares for a patient she does not think through every life's experience she has ever had which has made her what she is and therefore lies behind her ability to carry out that particular act of caring. However this does not touch the argument that the nurse's tacit knowledge of her expertise consists of her personal and experiential knowledge, and every act she carries out in her caring will come from her 'personal' and 'experiential' knowledge that is appropriate for that particular situation.

The essential role of practice of the formal reality for it to have meaning and be understood, challenges the concept of a rule being meaningful as a universal absolute. This is so because rules cannot be given meaning simply by other rules, or concepts by other concepts. The nature of the meaning of a rule or a concept lies in the way a rule is used and a concept applied, hence rules become *rule-following* (Johannessen, 1988), that is, the practices of the person, which entail the tacit knowledge of the person. Hence procedures which are abstracted from the 'declarative' expression of an expert using knowledge acquisition techniques, cannot be meaningful to the expert system as it can only process them, not practice them in a contextual reality as a human expert is able to do (with their formal or 'declarative' language of skills). This places a serious query as to the credibility of so-called decision making processes of knowledge based systems. Also, the procedures a system comes up with are only meaningful to the user in human-machine communication if the person knows how to practice them. It is increasingly becoming recognised that the initial idea of designing expert systems for naive users is not working, and that intelligent knowledge based systems are more effective if experts use them. However, as shown in case examples of doctors using expert systems, there are serious problems as the decision making process also involves the crucial role of dialogue with colleagues for the sharing of experiences and beliefs.

The dynamics of tacit and propositional knowledge means that they do not have static meaning. The relationship is almost dialectical, in the loose sense of the word. Yet 'knowledge engineering' reduces and fragments the nature of skill and its acquisition to the technicalities of getting data and processing it. One of the major crises hitting huge organisations which have attempted to use intelligent knowledge based systems, is the dilemma that the nature of the knowledge in these systems is static. The weakness of processing power and inefficiency of constantly updating these systems with the multitude of rapid changes in reality, gives rise to a huge bottleneck in 'information flow' in different parts within an organisation. This problem arises because there has been is an abstraction and reduction of the human communication 'system' which is highly complex, and the reduced system is unable to effectively deal with the change and uncertainty. The tacit dimension of human communication has been ignored because it is correctly thought to be too complex and not fully understood enough, for placing it within the system. One possible alternative to achieve a more effective kind of communication system is to take account of this tacit dimension. Instead of designing it from a machine-centred perspective (how to put the tacit knowledge into the system, and replace it with an information processing system), it should be designed from the perspective of identifying needs and purpose of applying the system within the context of the present human communications network within an organisation.

The issue of tacit knowledge is fundamental to an understanding of skill. This includes *performance* of skills. Just as the meaning of a concept or a rule can only be gained from its practice (Johannessen 1988), so can performance only be understood by what has gone to make it a particular performance. The Scandinavians (eg Perby 1988, Gullers 1988, Josefson, 1987) have studied in depth the nature of craftsmanship and tacit knowledge in nursing, photography, surgical instrument making, and weather forecasting, along with many other case studies. The focus is not on representation (of knowledge, as in knowledge engineering) but on understanding. 'To understand an utterance is to understand the person who makes it' (Molander 1989). Likewise with performance.

Performance of skills is a learned ability. Take the case of the Japanese dance craft of 'Waza' (Kumiko Ikuta 1989). This has a language of its own which describes and prescribes the process of development of the dancer. There is a distinction made between action (*Katachi*), and 'habitus' (*Kata*). The ultimate goal of achieving 'Waza', the mastered performance of the dance, is through a special metaphorical language ('craft language'). The role of this metaphorical 'craft language' is of central importance. It activates the learner's creative imagination, through which the dancer gets to learn her own 'Kata'. The metaphor for 'learning' is 'stealing in secret'. 'Katatchi' is 'an apparent physical form of action shown by the performer of a certain 'Waza', which may be decomposed into parts and described as a sequence of procedures'. However, 'Kata' is a culture laden form of action; 'it is gained not only from imitating an action independent of context, but is 'mastered' through committing oneself to or indwelling inside a certain culture or situation, therefore getting to grasp the situational meaning of 'Katachi' with a sense of reality.' (Ikutu 1989). The master transmits Kata to the learner by use of the metaphorical language using such phrases as eg 'hold your hand up as if you were trying to catch snow flakes', as opposed to saying, 'hold the hand at 45 degrees'. The language is 'action directed', a 'craft language'. The focus is on teaching with hints rather than details of Katachi to provoke sensation in the body of the learner to enable discovery of form (which is to be mastered). The metaphors are both comparative and interactive. There is a great deal more that belies the apparent behaviour in the performance of skill which cannot be reduced to a mechanistic/materialist process of behaviour, from objective observation, or model of mind.

In response to the behaviourists eg Watson and Skinner, or any 'scientific' account of human behaviour, Wittgenstein is cited, by the SCWL (Swedish Centre for Working Life), in his insistence that there is a problem of meaning with respect to human behaviour. He insisted that to those who do not understand specific practices, human behaviour although seemingly observable, will simply be unintelligible. Wittgenstein regards language as constitutive of human behaviour and that implies that we can never get outside of it to observe it.

In contrast with this framework of tacit knowledge and dialogue, AI is a world of models (Simon 1969) which are by definition closed, universal, and determinant with set parameters, variables and defined space. They are not just the abstraction of reality but are also its fragmentation. Fragmentation is apparent in the manner of the distinction made between 'knowledge acquisition' and 'knowledge elicitation'. The use of the metaphors, 'acquisition', 'engineering' and 'elicitation' make the assumption that the nature of skill can be reduced to structured 'data', 'information' and processes making it 'operational'.

A case study which I carried out on the nature of underwriting skills (see below) shows how taking account of the nature of the relationship between tacit knowledge and dialogue can achieve a fuller understanding of expertise because of it's participatory, sharing nature, compared with a 'knowledge engineering' approach.

Case Study: Skills of Underwriters
Purpose and Dialogue

This case study I carried out showed how the purpose of communication affects the nature of dialogue, involving eg appearance, body language, style, humour, perceptions, and the relationships of its participants established during it.

It involved a comparison of a dialogical participatory approach with that of communication using knowledge acquisition techniques used by a team of knowledge engineers who had already designed an expert system to identify particular risks in life insurance proposal forms. The systems design team decided to design a modularised system. Hence the purpose of the knowledge engineers' communication with the underwriters was to get the data from them under various categories (of insurance form sections), through devising 'techniques to elicit data or material' and then have a procedural method of processing this information. It becomes an abstraction of reality without any account of the human context i.e. personal experiences, culture, beliefs, values, etc. This was shown to be a great limitation for understanding the underwriters skills.

The purpose of my dialogue with the underwriters was to gain an understanding of their skills and expertise, with no predetermined presentation of what I understood them to be. I was the novice. This meant that they could talk of building up a picture of a person's way of life, movements etc, without the inhibition that this is the wrong or incompatible kind of information for the computer. In fact, as the dialogue progressed, they pointed out that a problem with representing knowledge in a static knowledge base is that static systems cannot keep up with these vagaries of world affairs - in particular the changing political climate within and between nations-, nor those that occur in changes in general life mores, values, behaviours, and so on, in societies.

The dialogue also indicated the extent to which the underwriters' personal beliefs and values come into making a judgement: their personal knowledge, and the tacit dimension of their decision making. People do, after all, make judgements which reflect their individuality and their culture. Rapport or empathy was established right at the beginning as tension inhibits good dialogue which needs joint participation.

I gave the underwriters a sample life insurance proposal form for reflection[1] and I informed them about the general objective which was to establish how they made a judgment about the applicant's riskworthiness. What was revealed was how the applicant's motive's and lifestyles are analysed, since the applicant's chances of living until a policy matures is critical. One of the underwriters stated that the information is only meaningful when it is '*seen as* a complete picture of a person'. The significance of this process of building up a picture of a person was that it resulted from the nature of the dialogue. It had not been achieved with the use of the knowledge acquisition techniques which had constrained the communication.

One of the most effective ways for the underwriters to explain why they were making a particular judgement was to tell a story or anecdote about a previous case from his own personal experience. The style in which these anecdotes are expressed will depend on the nature of the social communication. Because here, the underwriter was relating to me as a person - for example what values, beliefs, and humour the underwriter shared with me, he adapted his expressions accordingly.

This case showed how a focus on tacit knowledge ('personal' knowledge - socio-cultural life's experiences, and 'experiential' knowledge - working life environment) and participatory dialogue made for a significantly deeper understanding of the nature of the practice of underwriting skills. This is also illustrated through the following case study undertaken of consultancy practice.

Language of Practice - Consultancy Skills

Consultancy is about people communicating knowledge. What became clear from this study of consultancy skills was the the importance of cultural understanding and communication skills, and how this applies to even internal company communication channels, whether in a particular part of the company, in a particular country, or internationally, as well as when dealing with clients. It is clear that there is no one way to describe a process of consultancy. Different people have different methods and processes developed from previous experiences in dealing with clients and other companies. There is a richness in this diversity which makes for creative work. This is not the sort of skill or knowledge that can be abstracted into any procedural format using some of the afore-mentioned knowledge acquisition techniques.

What the study focussed on was the transfer of their experience amongst consultants themselves, that is the transfer of their tacit dimension of knowledge. With the consultants, as with the underwriters in the above study, there was a great deal of use of examples and stories. One example showing the need for cultural understanding as part of the job, by a consultant, was of his arriving in South America with his wife and being taken by his business host to the latter's family abode. There the consultant was treated very warmly by all the family members and friends. This was a cultural shock for the consultant as British cultural practices are very different, and the professional culture which, of course, reflects these. The consultant was confused as to how to show his gratitude for this unexpected yet friendly and generous welcome. Cultural understanding is also essential for the success of undertaking international deals, especially in an increasingly competitive world market.

Examples also make for dialogue, as others can relate their experiences and pool problems and possible approaches for solving them. Hence, examples can also act as a kind of *preventative* medicine. They can be checks which consultants can identify with and junior consultants can learn from. Metaphors (both visual and verbal) and catch phrases were also heavily used. These served to keep things in the memory and become part of the language of the consultants. They would mould the metaphors and phrases to explain different situations depending on the way they had interpreted them. Examples, anecdotes, stories, and metaphors are the richest forms of expression for relating experience as they allow for creative interpretation by the listener and a process of learning and understanding to tacitly occur (Gill, SP 1989)

The language the consultants used to both describe their own ways of doing things was a mixture of company-culture driven terminology embedded with company values, and their own personal language embedded with their own personal values. This seemed to be quite balanced amongst these consultants, who were very senior and whose years of experience both in this company and in previous other companies had probably enabled them to take up expressions which they felt comfortable with. Some examples of their self-description were: manipulating, devious, goals, truth, arrogance, interface, self-sell. Adjectives and values of conduct. The use of the adjective 'arrogance' is related to the pleasure many consultants took at being performers. *Truth* varied with consultants and contexts, and was regarded as problematic when discussing the element of salesmanship, which has some importance as seen in the heavy use of *self-sell*. Some saw their relationship to their clients as being that of an *interface* which to a large extent contradicts the notion of salesmanship and adds a different interpretation to self-sell. It was largely accepted that it is important to be 'manipulating' and 'devious' to achieve 'goals' which must be clear and defined.

These senior consultants were unanimous in their belief that 'a good consultant is someone who feels comfortable to know that there is a great deal that they still do not know'. This is particularly important for consultancy practice as it is 'largely about dealing with people and people are infinite - meeting a new person can sometimes change the previous framework of communication and work'. They were adamant that it is not possible to express their practice in terms of rules which could be used to understand situations. One of the problems students have on joining the company is that their formal education has trained them to think in a certain way, of applying formal analysis using certain clear methods. However, consultants have to use their 'intuition', 'imagination', 'creativity', 'think on the spot', which requires experience of practice. These skills cannot be transmitted through formal rules and set methods. The language of consultancy practice, in all its forms, becomes essential for learning and sharing experiences.

Conclusions

The underwriter case study shows that merely taking note of the propositional knowledge of the underwriters as categories of material data, did not make for an understanding of their knowledge and skills. The consultancy study showed the deep relationship between experience, practice and their language of communicating that experience in a non-propositional form. It highlighted the importance of ethics, feelings, cultural understanding, motives, purpose and empathy, all of which makes for the tacit dimension of their consultancy knowledge. It demonstrated the significance of the use of examples, metaphors both visual and verbal, by the consultants to express this tacit dimension.

The case studies highlighted the need for an alternative approach which focuses on the human user's tacit knowledge and practices. It is not a matter of considering the 'cognitive abilities' of an expert as an abstraction of the human mind, but a matter of understanding the practice of expertise of a person in the context of the working life environment.

It is proposed that such an approach might be developed through an analysis of tacit knowledge, through a study of the interplay of the forms of expression of knowledge in dialogue.

References

Alvey Annual Report (1987) G&B Litho Limited, London

Boden, M (1989) *Artificial Intelligence in Psychology: Interdisciplinary Essays*. MIT Press.

Berry, D (1987) The Problem of Implicit Knowledge. In *Expert Systems Journal.*

Cooley, M (1987) Human Centred Systems: An Urgent Problem For Systems Designers. In *AI & Society* Vol.1 No.1, Springer-Verlag

Cooley, M (1987) *Architect or Bee?: The Human Price of Technology* (New Extended Edition), The Hogarth Press, London

Diaper, D (ed) (1989) *Knowledge Elicitation: Principles, Techniques and Applications*. Ellis Horwood.

Ehn, P (1988) *Work-Oriented Design of Computer Artifacts,* Arbetslivscentrum, Stockholm

Fann, K T (1969) *Wittgenstein's Conception of Philosophy*. Blackwell, Oxford.

Gammack, J (1987), in Kidd (ed.)

Gill, K S (1986) Knowledge Based Machine: Issues of Knowledge Transfer. In *Artificial Intelligence For Society* (Gill, KS, Ed.), Wiley & Sons Ltd, Chichester, UK

Gill, K S (1988) Artificial Intelligence and Social Action: Education and Training. In Göranzon and Josefson (eds.) *Knowledge, Skill and Artificial Intelligence*, Springer-Verlag

Gill, K S (1989) Reflections on Participatory Design. In *AI & Society*, forthcoming.

Göranzon, B (1987) The Practice of the Use of Computers. In *AI & Society* Vol.1 No.1, Springer-Verlag

Göranzon, B and Josefson, I (Eds. 1988) *Knowledge, Skill and Artificial Intelligence,* Springer-Verlag

Gullers, P (1988) Automation-Skill-Practice. In Göranzon and Josefson (eds) *Knowledge, Skill and Artificial Intelligence,* Springer-Verlag

Guy, K (1987) *A Review of the Alvey Intelligent Knowledge-Based Systems (IKBS) Programme*, Science Policy Research Unit (SPRU), Sussex University

Greenwell, M (ed) (1988) *Knowledge Engineering for Expert Systems*. Ellis Horwood

Heritage, J (1984) *Garfinkel and Ethnomethodology*, Polity, Cambridge

Ikuta, K (1990) The Role of 'Craft Language' in Learning 'Waza'. In *AI & Society*, Vol 4.2.

Johannessen, K (1987) *Tacit Knowledge, A Research Report,* Swedish Centre For Working Life, Box 5606, S-11486, Stockholm, Sweden

Johannessen, K (1986) *The Centrality of Practice in Later Wittgenstein.*

Johannessen, K (1988) Rule-Following and Tacit Knowledge. In *AI & Society* Vol.2.4 , Springer-Verlag, London.

Josefson, I (1987) The Nurse as an Engineer. In *AI & Society* Journal, Vol.1. No. 2,

Josefson, I (1990) Language and Experience. In Göranzon & Florin (1990) *Artificial Intelligence, Culture and Language: On Education and Work*, Springer-Verlag.

Kidd, A (ed) (1987) *Knowledge Acquisition for Expert Systems: A Practical Handbook*. Plenum Press

Molander, B (1990) Socratic Dialogue: On Dialogue and Discussion in the Formation of Knowledge. In Göranzon & Florin (1990) (eds) *Artificial Intelligence, Culture and Language: On Education and Work,* Springer-Verlag.

Ostberg, O & Whitaker, R (1988) Channelling Knowledge: Expert Systems as Communications Media. In *AI & Society* Vol.2 No.3

Perby, M (1988) Computerization and Skill in Local Weather Fore-casting. In *Knowledge, Skill and Artificial Intelligence*, Springer-Veralg.

Polanyi, M (1966) *The Tacit Dimension,* Doubleday, New York

Potter, J & Wetherell, M (1987) *Discourse Analysis and Social Psychology,* Sage

Rosenbrock, H H (1988) Engineering as an Art. In *AI & Society* Journal Vol.2 No. 4

Simon, H A (1981) *The Sciences of the Artificial,* The Massachusetts Institute of Technology

Simon, H A (1976) *Administrative Behavior,* The Free Press, New York

Taylor, F W (1911) *Principles of Scientific Management*

Winograd, T and Flores F (1986) *Understanding Computers and Cognition: A New Foundation for Design,* Ablex Publishing Corporation, New Jersey.

Wittgenstein, L (1963) *Philosophical Investigations,* Blackwell, Oxford

Wittgenstein, L (1969) *On Certainty.* Blackwell, Oxford

Notes

[1]Gammack, J. & Morgan, S. {Bristol Polytechnic Project 1988-1989} came up with an idea of getting underwriters to identify combinations of categories of information on the insurance application forms, as a means of extracting the underwriters' knowledge. I used this idea to extend this work further and developed a dialogical approach. These category ideas are also referred to by Gammack et al., 'A knowledge acquisition and representation scheme for constraint based and parallel systems' in Proc. IEEE Conference on Systems, Man and Cybernetics, Vol. III, Cambridge Massachusetts, 1030-1035.

Applying Human Centred Concepts to the Development of Expert Systems in Manufacturing

Peter Holden

Introduction

The purpose of the workshop has been to establish an agenda for change; to devise practical means of introducing human centred concepts and approaches in the organisation, whether it relates to expert systems, computer integrated manufacturing or any other new technology. The principal concern is that technology should be used in a way which complements and stimulates human creativity, initiative and imagination rather than attempt to suppress these instincts for the misguided reasons of scientific rationalism or business efficiency.

Many businesses are aware of the importance of human issues. However, within the organisation, the mechanisms, communication channels and resources are not available to implement human goals. As a consequence, machine centred systems may arise by default, possible unintentionally, because a more proactive human oriented approach was not adopted in an otherwise technologically deterministic development process. It is therefore necessary to depart from tacit understandings of what human centredness means in the development of projects, and make explicit a conceptual framework from which human centred guide-lines can be defined and applied to the development lifecycle.

There can be no single definition of human centredness because there is no normative or collective model of human behaviour which accurately represents the role of the individual within the workplace. Rather than attempt to pre-define 'human centredness' and allocate social roles for the individual, the individual should be provided with the mechanisms so that he or she is able to make a social as well as an economic choice.

There are a number of reasons for the discrepancies in the definition of 'human centredness' stemming from the application of human centred concepts, rather than the principles upon which these concepts are based. First, the divergence of cultures and interpretations of human needs characteristic to a society or organisation. Secondly there is discordance in the perceived role of technology, and the problems to which it is applied, according to the organisational role of the individual. Depending on the level at which the individual operates - strategic, managerial, administrative and shop floor for example, there will be different levels of human centredness corresponding to the different tools and mechanisms used. For this reason, it is important to concentrate efforts on promoting human centred design at the conceptual level rather than at lower levels of analysis such as engineering or planning. In this way, human centred criteria may be applied to constrain the formulation and interpretation of company problems and therefore the way management views solutions to these problems and subsequently allocates technologies in the organisation.

The significance of human centred design is that the individual should be allowed to exercise and actualise personal goals as well as meeting objectives of performance and economy required by the company. There is no reason why the two perspectives, business and social should not co-exist. By developing technology in a way which accommodates human initiative, skill and other uniquely human qualities in planning, design and execution there is a more constructive and rewarding utilization of labour potential offering improved economic performance in addition to more desirable social consequences.

An important aspect of the human centred approach is to consider, with each successive technology, precisely what humans are being asked to accept. Within an implementation bias, this means honest and open demonstrations of capabilities and limitations. It should be made explicit whether the technology is intended to automate or to assist in aspects of the user's job, both as a business and productive entity and as a human social system. To enact a human centred focus requires a framework of evaluation which is deeply embedded in the organisational culture and is strongly present in the implementation of business goals and strategies. Human centred concepts should not be considered as an element of design, but rather as a process of defining business and organisational problems in human terms and allocating technological functions which allow humans to solve these problems. This requires that human centred objectives should be adopted by the organisation and become a formal part of the assessment process, not in a formal mechanistic sense, but in the form of human centred guidelines for all stages of development.

The Predominance of a Scientific Culture in Manufacturing: A Call for Change

The last few years have seen a visible transition from mechanisation to more flexible production systems brought about by changes in market requirements and expedited by advances in information and manufacturing technology. However, such innovations have failed to bring about any fundamental alteration to working practices and work organisation, other than possibly accentuate the dichotomy between the design, sanction and execution of physical and metal tasks. As Goldthorpe et al observed, although the technology may change, the basic social relationships will remain the same:

> *A factory worker .... can work at a control panel rather than an assembly line without changing his subordinate position in the organisation of production (Goldthorpe et al, 1971, p162)*

This is not a Luddite statement against the use of technology because technology has a central role in the self expression of human skills and creativity in the workplace. However it is a testament to the way in which technology has been used in manufacturing to sustain a machine centred and mechanistic model of the organisation which perpetuates management-worker relationships indicative of Taylorism. Within this culture, the scientific paradigm predominates in the selection and utilization of technology (Hoos, 1979). Technology has been allowed to redefine human capabilities using technical rather than measures based upon human characteristics. Even at the business level, a company's strategy may be defined as a series of information or machine technology projects, justified by the implicit belief that technological innovation alone will be rewarded with commercial success.

Advances in technology have diminished the divide between functions which can be automated and those functions which should be undertaken by humans. Arguments have focused on whether it is technically feasible to automate a function rather than whether it is socially desirable to do so. Managerial perceptions and preferences clearly play an essential role in this decision. Clegg (1988) speaks of how, within a scientific culture, the human component is judged as being error-prone and unpredictable and therefore to automate functions will improve the efficiency and certainty of the operation. This ethos is evident in the marketing and transfer of new technologies, particularly expert systems.

An expression of this ethos is the diminution of 'social choice' that the worker is at liberty to exercise in the workplace. For humans to be responsive and make use of their abilities, it requires that they have a noticable degree of autonomy and freedom to exercise any form of judgement in their work. However, when workers are subject to a pre-programmed series of tasks, often controlled and dictated by machines, there is little or no scope in the definition of the job to exercise autonomy. The personal effects upon the worker are that he or she is denied the opportunity to contribute towards the work outcome, with the concomitant effects, as Reed (1987) notes, that *'creativity, intuition, confidence and imagination become out of practice'*. This serves to fuel the self-perpetuating and self-validating arguments for Taylorism (Clegg & Dunkerly, 1980). The 'adverse' human reaction to specialised, repetitive work will confirm management feelings that tight control over shop floor workers and work practices is necessary to produce goods and services efficiently. The responsibility for conceiving, planning and initiating work tasks is therefore retained by management, with the effect that in many manufacturing and engineering companies today, the design, planning, and transactional functions are larger than the manufacturing function itself.

In order to effect change, it is important to recognise the failures of the old regime. The UK and other European countries have been held responsible for failing to acknowledge these failings (Matsushita, 1989). Indeed, perhaps it is because Taylorism and the wider scientific paradigm is so deeply rooted in our attitudes towards work, education and training that we are now not aware of its presence in the same way that a person is not aware of being asleep?

Whatever the reasons for the perpetuation of the existing scientific culture, it is becoming increasingly apparent that the effectiveness of projects on advanced information technology, such as Computer Integrated Manufacturing (CIM) and expert systems, lies not in the technology alone but as organisational systems with the determinants of success being the qualities of the humans which the inherent machine centred focus tries to displace (Ravden et al, 1987). This is highlighted by the work of Brödner (1986) who showed how Computer-Aided Process Planning (CAPP) systems developed in West Germany, allowed the engineer's skills and deep knowledge of the production process to "waste-away". Kelly (1989) too, describes the failures of automated operations in datacentres. Already, with leading-edge technologies such as expert systems, there are examples of technological misuse. Östberg (1988) for instance, comments on how expert systems have begun to use Taylorite measurement techniques to codify knowledge; Jones (1988) voices concern over the deterministic nature of the expert systems market; and Mumford (1987) describes a knowledge replacement focus in the development of a notable expert system - indeed this is actively encouraged by some (see for example Feigenbaum (1988)). All these examples are testament to the failure by industry to recognise the essential limitations of technology. They also reflect the extent to which informal and subjective criteria have been rejected in the development process, leading to the widespread dismissal of social criteria in the evaluation process.

Towards an Empirical Understanding of Human Centredness

The term 'human centred' suffers the same fate as does 'expert systems', in meaning different things to different groups of people. This reflects the variety of cultures and interpretations of human need characteristic to a society or organisation. However, within the range of definitions, lies a number of universal human centred objectives which transcend cultural and technological variations. Discrepencies lie in interpretation and the mechanisms with which to achieve human centred goals rather than the goals themselves: as a developing approach, attempts should therefore be made to consolidate and strengthen these universal concepts rather than emphasise their differences.

Although labelled 'human centred', it would be wrong to consider these concepts as applicable solely to problems and issues which arise at the personal level. Taylorism, for example, was not a model of individual behaviour alone, but a conceptual approach which has influenced work organisation and management practice. In defining alternatives to Taylorism, it is necessary to search beyond the individual setting to consider more societal and structural settings. For 'human centred concepts' to be effective therefore, they should be viewed not as a singular discipline operating at the microscopic level, but as a multiple-perspective concept pertinent to all levels of analysis.

By studying the role of the individual from these different perspectives and levels in the organisation, the orientation of the individual becomes an important issue. The individual will be influenced by the economic, social and even technical dimensions of the business which will constrain particular forms of behaviour and promote others. The individual exhibits 'emergent' characteristics (Markus & Robety, 1988) which are contingent upon these dimensions. It is simplistic therefore to define unidirectional processes of deskilling and motivation within the structure of human centredness. These make assumptions about individual behaviour at the general levels such as *all individuals require self-expression*, or *must exercise responsibility*.

A pilot project was undertaken in the client organisation to develop an expert system for capital investment appraisal, intended to assist with the data-collection and routine decision-making, tasks of financial accountants. The system failed, despite initial support, because the accountants decided that the routine aspects of their job added stability and security as a necessary complement to the more creative and knowledge intensive elements of their occupation. We must not presume to allocate social roles for the individual. This paternalistic role is as misguided as Taylorism in its simplistic intentions of an 'aggregated' human being. Rather the individual should be given the social option to choose the level of automation, the level of routineness or the degree of responsibility, and that visible mechanisms are made available within the organisation to make this possible.

The popular interpretation of human centred concepts appears to apply at the design level only. In a manufacturing context at least, 'design' presupposes technology as the solution and as such, the scope of human centred concepts is limited to discussions on the allocation of functions between humans and machines. Human centredness should be interpreted as a process of participative, incremental, change, rather than a particular type of technology embodying specific social principles. If the aim of human centred concepts is to accommodate social choice and human need in the development and operations of new technology, then it should focus not on the technology but on the problem situation and how it is understood. As such, human centred design should encompass earlier stages of problem conceptualization and requirements analysis so that the problem is expressed in terms of human as well as technical needs.

Defining the Level of Human centredness

One of the main problems of introducing a programme of human centred change is that the concepts will appear novel and unorthodox. Many large firms retain traditional working practices and mechanistic structures for reasons of stability. Rooted within these conventions is a particular company culture and political structure which are inflexible and resistant to organisational changes. In the client company, for example, upper management was quite enthusiastic about adopting a human centred approach, but the strata of middle management were reluctant to participate. This was because the middle managers had resided in the company for some years and had often been promoted from the shop floor. By contrast, senior management tended to be fairly young, university trained and had experience of working in a number of organisations and therefore tended to be more flexible about change.

The organisation is not homogeneous, and at each level there will be different cultural and political barriers to contend with when implementing human centred concepts. Sensitivity is therefore required in defining the appropriate scope and entry level, and deciding upon the most appropriate medium with which to communicate these concepts. The level at which most documented human centred projects appear to concentrate is at the shop-floor level, primarily relating to machine technology. Human centred numerically controlled machines, for example, are identified with craft operators performing programming tasks. In Japan, this is associated with interactive programming and production island studies at Hitachi and autonomous distributed Flexible Manufacturing Cell (FMC) at Keizai Shinbun.

Human centred concepts are not exclusive to this domain though. Where technology has redefined social conditions, at whatever level of the business, human centred concepts are equally applicable. An automation focus can exist in an administrative as well as a production context. For example, in large organisations there is a myriad of white-collar workers whose jobs are being automated or demeaned. The scope for a human centred approach should also be extended into managerial functions too, because new technologies have begun to redefine management role in terms of structure, decision-making, interpretation and control. For example, Wells (1987) speaks of the displacement of the traditional design experiences in developing expert systems in a client organisation have shown that professionals and engineers are concerned about losing their "feel" and intuitive skills to these systems. Indeed a demonstrator expert system for personnel selection was abandoned because management perceived a loss of control over established manual recruitment procedures. With the development of *knowledge based decision and information systems*, the traditional role of management to interpret and restructure information is being automated with precisely the same personal effects as experienced by manual workers.

Addressing Managerial Perceptions

Management in industry are often adverse to dealing with the non-specific and the non-measurable such as attitudes, values, feelings and behaviours. They often fail to understand in practical terms what it is necessary to establish a new culture or management style and the substantial tangible and intangible benefits that arise from such an approach. Indeed early experience at implementing human centred concepts (Esprit, BMFT, for example) have shown the difficulties in overcoming managerial resistance. By addressing managerial perceptions (fears, prejudices and misunderstandings) within an effective conceptual framework, it is possible, I propose, to achieve a more lasting and pervasive commitment towards human centred concepts throughout the organisation. Management are inherently sceptical of new methodologies and tools and are often capable of deciding upon the right mechanisms for human centred change in the organisation, once they are convinced of the value of these concepts.

Sell (1987) has mentioned the importance of management perceptions in the development of expert systems. For it is at the decision making level that the motives for innovating and the perceptions of the role of technology by management may be influenced by machine centred arguments (Clegg & Dunkerly, 1980). The machine centred focus is a function of how management perceive the value of human 'worth'. To change these values requires a change in viewpoint and understanding of the problem.

It is at this level of abstraction that the human centred approach will have the most impact. By expressing business and organisational problems in many different ways, attention is drawn towards the different contexts and needs - business, organisational, behavioural and technical for instance - and more particularly, how each of these contexts interact and interrelate. A useful conceptual approach which highlights this interdependancy is the Multiple Perspective Concept (MPC). This was developed specifically for technology assessment (Linstone, 1981) but that can be used to help managers understand and structure problems from more than the single, serial, technical dimension. MPC communicates to management that not everything can be expressed in rational and quantitative terms and that the informal dimensions - political, cultural and behavioural needs for instance - are of equal importance. In terms of expert systems development, MPC promotes a symbiosis between the formal and the informal which reflects the interdependancy between explicit and tacit forms of knowledge.

There are three primary perspectives in the MPC framework (Fig1(enclosed)): the technological (T), the organisational (O) and the personal/individual (P). There are also numerous settings according to the interface between each of these perspectives. However, the essence of MPC is that any perspective may illuminate any element. Although, for example, a 'T' perspective is necessary to understand the mechanics of expert systems, the 'O' and 'P' perspectives add important insights which amplify the developers' understanding of the problem.

The 'decision focus' represents the culmination of settings, specific to a problem (this is signified by the shaded area in Fig1). It helps management to identify a focus of perspectives, the point at which decision-making and assessment should be made. Clearly, this focus will change according to the nature of the problem. For structured problems, the dominant perspective is technological and the decision focus will shift to point A in Figure 1. For less structured problems, where the decision style is necessarily judicial and intuitive, the 'O' and 'P' perspectives play an important role in defining human needs, and the decision focus will shift to somewhere near point B in Figure 1. The 'O' and 'P' perspectives also help developers to see the limitations and failings of the T perspective and help redefine 'analysis' away from purely scientific endeavours into the realms of human centred, socio-technical systems.

The Significance of Human Centred Concepts to Expert Systems Development

The simplification of human skill into a number of highly specialised and repetitive tasks, which characterizes Taylorism, has removed workers discretion and their ability to exercise practical knowledge. It is this emphasis upon explicit knowledge and dehumanisation which may pervade in the development of expert systems (Holden, 1989). A consequence is that the level of control exercised by the user in the organisation is diminished in favour of machine control. Where mechanization has downgraded the physical tasks of the manual worker, expert systems may be viewed, erroneously, as a means of automating human cognitive processes. The implications for skills and use of knowledge are even more discouraging and alarmist and for this reason, the application of human centred concepts is of more importance than ever before.

The development of expert systems and knowledge based systems is part of a wider trend towards centralised information management systems. With this trend, information systems control becomes governed by globally optimised policies which, though technically valid, may be wholly incomprehensible in a human context. The situation may emerge where humans in any one part of the expert system cannot exert substantive control over that part of the system because they lack knowledge of the objectives and operation of the total design. Humans may therefore be made responsible for performing actions and taking decisions which they don not fully understand and do not have the confidence to question. This is particularly problematic in failure type situations because, in all cases of automation to date, humans have remained the ultimate backup system.

A human centred approach towards the development of expert system would not segregate knowledge, but segregate the allocation of functions in the organisation. This would reduce the scope of applied knowledge but still retain a complete mental activity in terms of presentation of skills, margin of autonomy and potential for the individual to learn more. By contrast, expert systems developed from a machine centred focus provide a static representation of human behaviour because there is no further scope for human involvement and intervention. Operators subordinate the sensory and actuating portions of their work to the machine: information is aggregated before the operator is allowed to make any decisions. It therefore becomes a variety reducing exercise where the operator is trained how to use the expert system rather than to learn the general principles and concepts behind the operation (Holden, 1990).

Human centred concepts provide a useful framework for defining basic social criteria which then drive the development process. When developing expert systems the human goals might be that:-

a) the user should have complete control over the expert system
b) the expert system should allow for the regenerative use of skills and knowledge
c) the expert system should be sensitive to the exercise of discretion by the user
d) the user should feel confident and be able to exercise discretion
e) the expert system should not replace the communication medium of free learning on a one-to-one basis
f) management should not try to retain a monopoly of knowledge

Rather than impose a regulatory emphasis however, expert systems technology should be viewed more positively as a means to achieve human centred goals. Klein (1988) for instance describes the potential of expert systems to restructure the organisation. Other possibilities for human development through the exploitation of expert systems might include:-

a) improved learning and training with new understanding emerging through experience with and use of knowledge based models
b) increased knowledge processing capacity
c) greater independence by reducing the reliance upon specialists and experts
d) improved personal effectiveness and efficiency due to faster and more consistent applications of routine knowledge
e) a deregulation of technology
f) an emphasis upon practical knowledge in training and education

These benefits will be contingent upon the personal needs of the user and of the business and organisational settings in which the user operates.

It is important to know where the boundaries for human centred concepts lie. There are areas in the organisation where it would be inappropriate to apply a human centred approach: for instance, for highly synchronised and co-ordinated tasks, such as data entry and data processing, the human role is simple, mundane and unfulfilling. It would be difficult and undesirable to use human centred arguments for sustaining human role in these areas. By contrast, for more creative and uncertain tasks, such as engineering design, it is imperative that human skills are maintained (Rosenbrock, 1988). Enclosed Figure 2.a shows that for a given level of task complexity, the 'net human worth' is equal to the perceived human return on adopting a human centred programme offset against the opportunity costs of not utilizing automated technology. At point X-X in figure 2.a, the net human worth is negative which suggests that the company should look towards automating the task in some way. As the task complexity increases and the limitations of technology become evident, the relative returns on human investment increase until at point Y-Y In figure 2.a, the net human worth is positive, indicating that a human centred approach would be worthwhile. The role of technology in this case is as 'human facilitator' (Cooley, 1988).

Chandrasekaren (1988) identified generic groups of tasks in the organisation which exhibited similar characteristics in terms of complexity, syntax, information requirements and skill requirements. On the basis of a similar classification, a company could undertake 'human audits' on functions of the business and decide where it is essential that human centred design should take place, or conversely, where automation is germane.

Figure 2.b shows that the related change in costs at a given level of task complexity. As complexity increases, the technology costs (capital, maintenance, support etc) escalate at a much higher rate than the direct (extrinsic) human costs - such as training costs, improvements to the working environment and other investments in the individual. It is significant that the indirect (intrinsic) human costs, arising from motivational, deskilling, and related problems for example, will generally decrease as the task complexity increases, thereby offsetting the direct human costs.

Implementing Human Centred Concepts: Addressing the Scope of Analysis

There are significant social, cultural and educational differences between countries and this will result in national differences in the strategic application of human centred concepts. A particular culture may appear to approximate 'human centredness' more than others. For example, Japan is now trending to introduce Flexible Manufacturing Systems (FMS) in order to improve quality rather than saving labour as in the U.K. There are also industrial practices characteristic to a nation which inhibit the transition to a human oriented development programme. In the UK for example, problems tend to centre around senior management undue emphasis upon financial control, and lower management emphasis upon regulation and functional control. By contrast, Japanese senior management, with their superior accounting systems and greater engineering knowledge at higher levels, are more receptive to human centred concepts and much of what lower management already do may be described as 'human centred'.

However, in making this comparison, we must be careful of adopting a Japanese model of human centredness. The application of Japanese industrial concepts in the UK, such as 'Toyotism' and production techniques such as 'Flexible specialisation' have perhaps contributed to an economic transformation, but the personal ramifications may be less desirable; - British supervisors at the Nissan factory in Washington (N.E. England) for instance, describe the brain-washing and indoctrination which they face. This is a reminder that human centred definitions should favour the side of the worker rather than that of capital.

The essence of human centred development is an eclectic approach towards the use of tools and methodologies. The choice of tool should be governed by the emergent t human centred properties of the organisation and used alongside other tools from within a common conceptual framework. There are numerous approaches which could be said to embody or facilitate human centred principles, and their use should be contingent upon the particular culture and settings of the organisation at any one time, also upon the structural level within the organisation at which they are expected to be used. There are two basic levels of application: the strategic or business level, and the development level. These will be discussed separately as follows.

The Use of Tools at the Business Level

Although human centredness should be applied to all aspects of the organisation, the mechanisms are clearly different. There are different levels of business decision making and the scope of the relevant technological decisions will therefore vary. For instance, an expert system which aims to control the operations of company information systems for example, will have strategic implications upon *business effectiveness* in terms of quality, accessibility, speed and accuracy of information internally, and also to suppliers and customers. The technology would transform nearly every individuals' system of work, either directly or indirectly, with consequent favourable or unfavourable personal effects. The scope of these changes requires that human centred concepts are applied at the organisational level - a policy statement on human centred design for example. By contrast, for an expert system constructed for instance to assist computer specialists solving occasional problems - as in the case of the client organisation - the business consequences are small and can be expressed in terms of improving the *operational efficiency* of a particular business function. In this case, the personal effects are clearly bounded and will apply at the individual rather than at the business level. It is more appropriate at this level to concentrate on providing human centred guidelines as part of a less formalised educational process, sensitive to the particular settings of the function or department.

By drawing attention to the likely business impact of the proposed technology therefore, it is possible to decide on the level and organisational context to which human centred concepts should be applied. It also highlights an earlier point that individuals have different organisational roles and each will necessitate a different human centred focus. A useful tool in this process is the Portfolio Matrix (Ward, 1986). This splits the business impact of new technology into four main categories, as shown in enclosed Figure 3. These are: *strategic, tactical, operations, and support*. We may add to this a human focus (shown in Fig3) which communicates the boundaries within which human centred design should function.

From work in the client organisation, approximately 200 possible expert systems projects were identified by users at all levels in the company, of which five were selected for development. It is significant that of the 200, nearly 50% were *support systems*, as shown in enclosed Figure 4. The value of expert systems was clearly seen by managers and staff in the company as a means of improving their own personal effectiveness in performing specific tasks. This suggested that, for greatest impact, the development of expert systems within a company should use human centred guidelines which were directed at the individual level, rather than at the organisational or strategic levels.

The Use of Tools During Development

There is no shortage of tools for the development of expert systems. However, those tools and techniques which are currently available have rational and quantitative orientation which leads to a particular type of technical evaluation. As Hirscheim et al (1987) comment:-

It is not unreasonable to state that assessments which are entirely objective and rational will prescribe 'scientific' solutions. Evaluation has been misdirected towards tools and techniques and way from understanding

The selective use of these tools can be useful and in some of the more technical aspects of design, are necessary. However the preoccupation with choice of tool may distract attention from the real issue of providing systems for the organisation which are useful.

Attempts have been made to incorporate human factors in development, but as a parallel activity to technical design and operating at the component level rather than serially, at the systems level. For example, Diaper (1988) has developed a 'people orientated' methodology for requirements analysis; Hart (1986) describes the prominent role for human factors in knowledge elicitation techniques; Kid (1985) mentions user-driven knowledge representation techniques; and Hammond al (1983) apply human factors to the design of the user interface. Using the tools at each of the development phases will likely produce more 'usable' expert systems, but their cumulative effect is not necessarily to produce a human centred system. Designing a human user interface, for instance, is of limited use if the expert system functions in a mechanistic work environment and the user is restricted to operating the system in a particular way. These tools apply human factors from within a formal technical perspective and concentrate their analysis at the level of technology rather than at that of the individual. Furthermore, there is no common human conceptual framework within which each of these tools operate so there may be conflict and 'dissonance' (Den Herlog, 1982), because formal tools are highly task oriented in their approach to human issues. Human factors are applied sub-optimally to achieve certain task-goals at a particular technical level and, as such, the integration of levels may lead to a conflict of goals. Human centred design, by contrast, should aim to be process oriented. The distinction lies in the importance attached to the development process itself, rather than just the deliverable. A human factors approach uses scientific and computational methods to allocate human functions within the technical system. Human centred design, by contrast, uses conceptual and informal processes to allow for the human use of these formal methods. By placing a human context in the use of these tools and expanding the scope of 'analysis' from beyond the technical interface to consider social and organisational interfaces and their interaction, the focus of the system is changed from being machine centred to human centred. As a result the problems practitioners describe of applying human factors in industry (Bright et al, 1989) are likely to be reduced. Techniques which promote a process oriented approach to development consider the technology as human social systems and therefore recognise the prominance of human and organisational issues as part of the analysis. Mumford's ETHICS approach to design by participation (Mumford, 1983); and Checkland (1981) and Wilson's (1984) use of Soft Systems methodologies for problem identification and systems design respectively are good examples of this approach.

Justifying a Human Centred Approach Towards Development

Human centred design is constrained by the formal investment appraisal techniques which are used in determining the business value of the new technologies such as expert systems. These measure the quantifiable benefits, without accounting for the 'intangible' and qualitative human benefits. Because these 'soft' issues are excluded from the analysis, it often results in an acceptable return on investment is not indicated and therefore the investment is not allowed to go ahead. Change is required not just in the way projects are evaluated, but also at the in the organisation in which the appraisal takes place. Primrose and Leonard (1987) suggest that investment appraisal should operate at the company-wide level rather than at the operations level. By doing so, the investment focus changes from a situation in which technology necessarily 'competes with humans' to one where human capabilities are complemented through an emphasis upon improving the competitive ability of the business. By changing the focus, the authors note that:

> *... a new range of applications emerge, which are potentially more profitable investments and do not exacerbate the problem of (human) acceptability. (Primrose & Leonard, 1987, p 252)*

Central to the discussion of appraisal, is the problem of how, in a commercial enterprise, the concept of human 'worth' and 'social value' is reconciled with the more tangible and aggregated concept of business value? Brödner (1988) has shown that human- centred approaches, although placing significant emphasis upon human worth can result in both human and economic benefits. The strengths of human creativity coupled with the innate ability to handle complexity, uncertainty and change, can be expressed in business terns. These include, amongst others possible benefits, the following:-

a) reduced overheads,
b) reduced project risk,
c) improved machine utilization,
d) reduced labour turnover and,
e) improved system capabilities.

However, the relationship between human potential and business return is not so straightforward because there will be an increase in costs through adopting a human centred approach. These represent the costs to the organisation for pledging a greater commitment to the individual, manifest through improved training and the likely reorganisation of the technology layout and work organisation. Returning then to Figure 2.a, the decision should be taken by management, whether, for a given level of task complexity, the benefits arising from a greater investment in the human role are greater than the opportunity costs of investing in purely automative technologies. Although there is a wider philosophical argument that human centredness should be an end goal in itself, without the need to justify it in business terms, it is likely that this approach will be the only possible means of introducing human centred concepts to manufacturing for some while.

Human centred criteria should become key factors at the strategic level, involved not only in the use of technology as a design issue, but in a more prescriptive capacity, constraining the appraisal of technology and the interpretation of technology policy. Not only can technology be regulated by human centred guidelines without sacrifice of its economic purpose, but by applying these guidelines in investment appraisal, the social and organisational costs will be made visible. This will serve to highlight the unsuitability of prosed technical projects, and to reveal the misinvestment of existing technologies which are presently in use in the organisation.

Conclusions

Taylorism is a manifestation of the scientific approach which evolved from a particular technical and organisational setting. Although these settings have changed, significantly in the technical domain, the underlying conditions which procreate Taylorism remain unchanged. The development of advanced manufacturing technologies, such as expert systems, and attitudes towards technology assessment generally, are implicitly deterministic. Individuals will have personal needs and aspirations, but these considerations are all too frequently marginalised by the pre-eminance attached to technical and commercial settings, despite the criticality of the individual focus in the development of expert systems. Even when management recognise the importance of individual needs, often the mechanism and communication channels required to achieve human objectives Are either unavailable or unattainable. Only by elevating human issues to the front-end of technology assessment and describing failure in industry in human terms can this machine centred focus be overcome.

The benefits of adopting a human centred approach can be communicated in business terms. The initial costs of attention to identification is offset by the increased likelihood that more usable systems will be produced. Furthermore, by introducing a process which illuminates the social and organisational costs, rather than simply the resource and capital costs, managers may become more selective and look beyond the simplicity of technical feasibility to consider the criteria of desirability, social impact and human need. IN prescribing expert system based solutions therefore, a business must measure the perceived benefits of developing an expert system in a certain way against the opportunity costs, when a human resource, the user, produces value below potential. Education, current skills and underlying social, cultural and psychological factors will all influence the net worth of the user and should therefore be considered in the assessment process.

There are a number of tools and methodologies which promote the inclusion of human factors in the design process, but very few which encourage the development of human centred systems. Techniques which may be said to embody the principles of 'human-centredness' combine formal and informal methods within an emergent conceptual framework for the organisation. They concentrate on the human processes of change as well as the end product, and consider technologies as human social systems operating at the interface of many perspectives rather than the singular technical perspective.

Acknowledgements

The author would like to thank Professor Masuda of Tokyo Keizai University, Dr Richard Badham of Berlin and other members of the Manufacturing Sub-Group at the International Workshop on Human centred Design.

References

Bright, C, Inman, A & Stammers, R (1989) Human factors in expert systems design: can lessons in the promotion of methods be learned from commercial DP? In *Interacting withy Computers* 1(2), August 1989, Butterworths.

Brödner, P (1986) Skill based manufacturing vs. 'unmanned Factory' which is superior? In *Int.Jrnl. of Industrial Ergonomics* 1(86)145-153, Elsevier Pub.

Chandraskaran, B (1988) Generic tasks as building blocks for knowledge based systems: the diagnosis and routine design examples. In *Knowledge Engineering Rev*. 1988.

Checkland, P (1981) *Systems thinking, systems practice.*Wiley

Clegg, C (1988) Appropriate technology for humans and organisations. In *J.Inf. Technol.* 3 3, 133-145.

Clegg, S & Dunkerly, D (1980) *Organisation, class and control,* Routledge & Kegan Paul, New York USA.

Cooley, M (1988) Creativity, skill and human centred systems. In Göranzon, B. & Josefson, I.(eds) *Knowledge, skill and artificial intelligence* Springer-Verlag, Berlin, 127-137.

Den Herlog, J F (1982) The role of information and control systems in the process of organisational renewal. In Lockett, M. and Spear, R (eds.) *Organisations as systems,* Oxford University Press, Oxford UK, 112.

Diaper, D (1988) The promise of POMESS: A people orientated methodology for expert system specification. In Berry, D & Hart, A (eds) *Proc Human & Organisational issues of expert systems* (Joint ICL and Ergonomics Society Conference) Stratford-Upon Avon, UK.

Feigenbaum, E A (1988) *The rise of the expert company* Macmillan Press, New York., USA.

Goldthorpe, J H & Lockwood, D (1971) *The affluent worker in the class structure,* Cambridge University Press

Hart, A (1986) *Knowledge acquisition for expert systems* Kogan Page London.

Hammond, N, Jørgensen, A, Maclean, Barnard, P & Long, J (1983) *Design practice and interface usability: evidence from interviews and designers* IBM Hursley Human Factors Report HF082 Hursley Park, Winchester, UK.

Hirschheim, R & Smithson, S (1987) Information systems evaluation: myth and reality? In *Information Analysis* (ed. Robert Galliers) Addison Wesley.

Holden, P (1989) Working-to-rules: A Case of Taylor-made expert systems. In *Interacting with computers* 1(2) August 1989, 197-219. Butterworths UK.

Holden, P (1990) Human centred Expert Systems: A response to Taylorism and the scientific paradigm. In *Interacting with Computers*, Vol.2(1) Butterworths UK (forthcoming)

Hoos, I R (1979) Societal aspects of technology assessment. In *Technological Forecasting and Societal Change* 13, 3, 191-202.

Jones P (1988) Showing some intelligence. In *Informatics*, March 1988, 9(3).

Kelly, J (1989) Nightmare in Dreamland. Cover story in *Informatics*, September 1989, 1(9) 36-40.

Kidd, A (1985) What do users ask? - Some thoughts on diagnostic advice. In Merry, M (ed) *Proc. Expert systems* 1985, Fifth Technical Conference on BCS Specialist Group on Expert Systems Cambridge University Press, UK.

Klein et.al. (1988) Organisational Structure: the implications of expert systems. In Berry, D. & Hart (eds) *Proc. Human & Organisational issues of Expert Systems* (joint ICL and Ergonomics Society Conference) Stratford-Upon-Avon.

Linstone, H A (1981) The multiple Perspective Concept. In *Technological Forecasting and Social Change* 20, 275-325.

Markus, M, Lynne & Robey, D (1988) Technology and organisational Change: causal structure in theory and research. In *Management Science* may 88 583-598.

Matsushita, K (1989) qouted by Professor M.Cooley in "Why our vision of the next century should be in a class of its own", Frontiers' in *the Guardian* Newspaper(data unknown): edited by Peter Large.

Mumford, E (1983) *Designing participatively* MBS publication.

Mumford, E (1987) *The successful design of expert systems-are means more important than ends?* paper presented at Oxford P.A.Templeton College, Oxford, UK (unpublished)

Ostberg, O (1988) Applying Expert systems technology: division of labour and division of knowledge. In Göranzon, B & Josefson I (eds.) *Knowledge, Skill and Artificial Intelligence*, pp169-183. Springer Verlag.

Primrose, P L & Leonard, R (19xx) An approach for negating the view that technology' competes with humans. In *Robotica*, 5, 251-255.

Ravden, S J, Clegg, C W & Corbett, J M (1987) *Report on Human Factors Criteria for CIM Systems & Methods to Enhance their usability*, Esprit CIM project No.534.

Reed, E S (1987) Artificial Intelligence or the mechanisation of work. In Open Forum in *AI & Society* 138-150 Springer Verlag.

Rosenbrock, H H (1988) Engineering as an art. In *AI & Society* 2(4) 315-320, Springer Verlag.

Sell, P (1987) Strategic issues in introducing knowledge based systems. In *KBS* 1987, Online publications, Pinner, UK p401.

Ward, J M (1986) Strategic Information (IS) Management in Information Management. In *Competitive Success*, ed. PM.Griffiths, Pergamon Press, 1986.

Wells, C S (1987) The design supervisor - a changing role with CAD? In *IMechE J.* C369/87 27-35

Wilson, B (1984) *Systems: Concepts, Methodologies and Applications*, John Wiley & Sons Chichester, UK

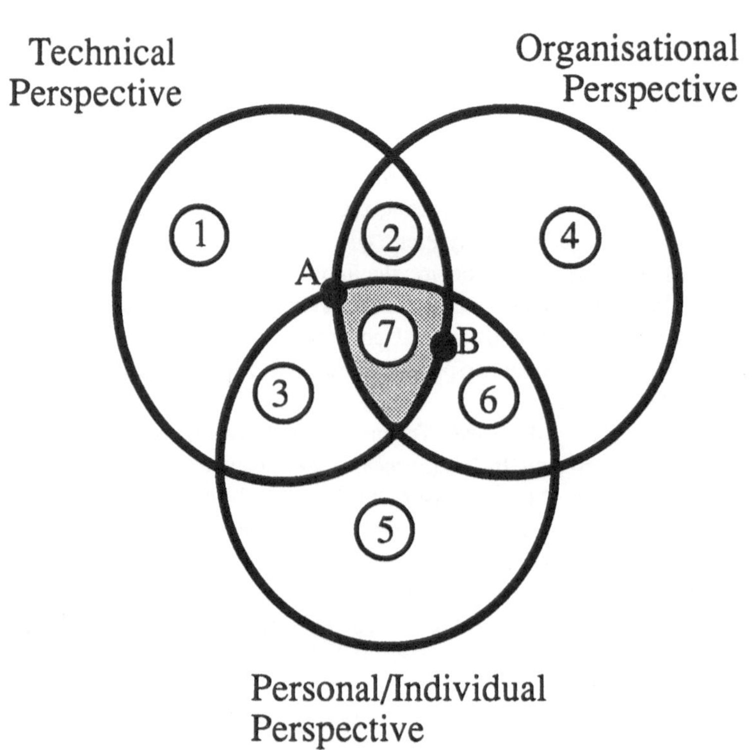

Settings:-

1 - Technology & Physical Environment
2 - Socio-Technical
3 - Techno-Personal
4 - Organisational
5 - Individual Actors
6 - Political Actions
7 - Decision Focus

Figure 1: A Diagrammatic Representation of the Multiple Perspective Concept

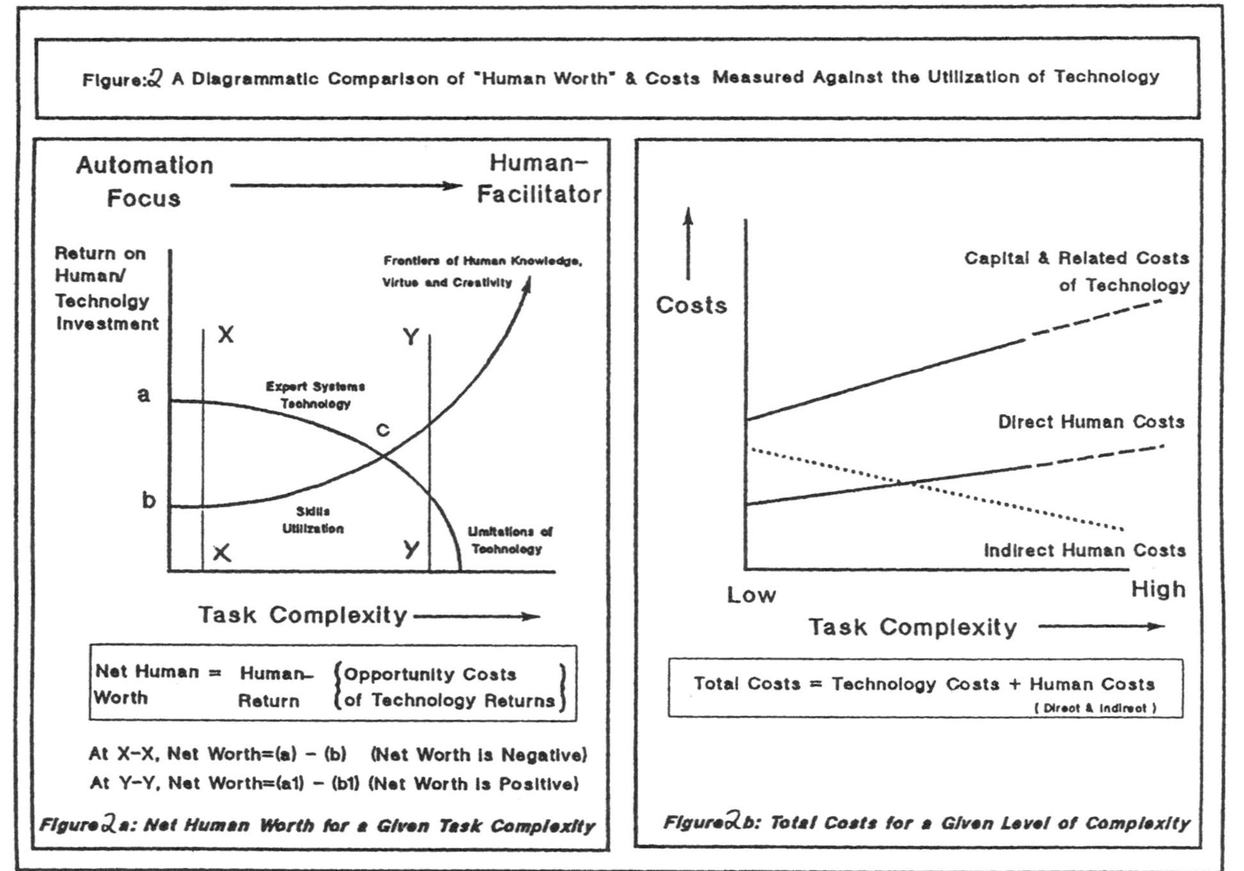

Figure 2a: Net Human Worth for a Given Task Complexity

Figure 2b: Total Costs for a Given Level of Complexity

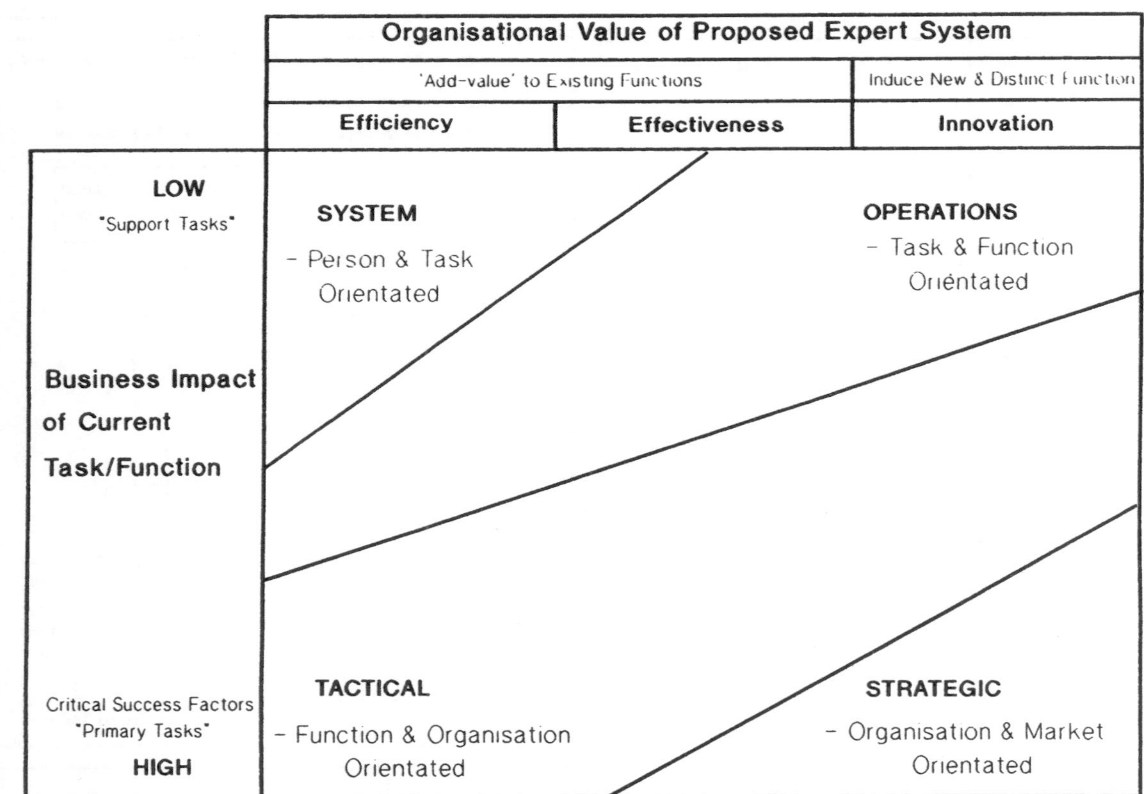

Figure 3: A Revised Framework Defining Regions of Business Impact & Organisational Value

Figure 4: A Classification of Expert Systems According to Perceived Business Impact

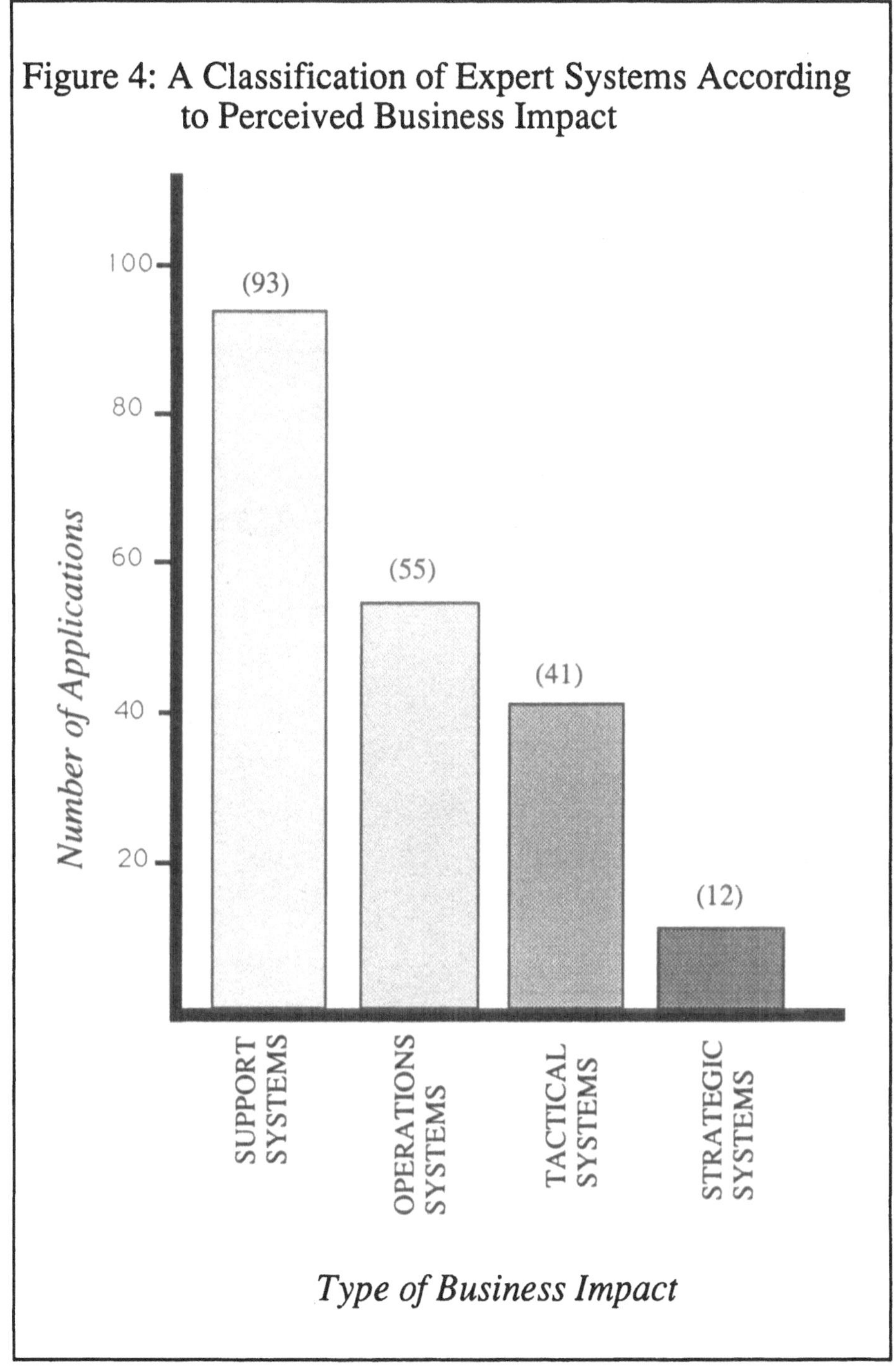

Anthropocentric Principles - Tools and Organisation of Industrial Design

Lauge Baungaard Rasmussen
Poul Tøttrup

The Anthropocentric Approach

In the recent decade a number of empirical studies (Cooley 1981, Finne 1982, Wingert 1984, Majchrzak 1987, Rasmussen 1987, 1989) indicate that the introduction of Computer Aided Design (CAD) systems may present problems in design activity similar to those experienced in some fields of skilled manual labour when subjected to technical change[1]:

- *a further division between creative and non-creative activities*

- *a further intensification of work*

- *an increased rate of design-knowledge obsolescence*

- *a decrease in contact and fewer opportunities for informal, face-to-face communication.*

In general, the strategy has been to subdivide and codify the designer's knowledge and subsequent reduce the design activity to a 'series of choices' in a formalized and standardized system of possible actions.

The Anthropocentric approach follows another path: the designers' work culture based experience and action are being reflected and priorised in a dynamic relation to new technological possibilities, instead of being subsumed or made obsolete.

According to the Anthropocentric Approach, the immediate re-investment of rationalization profits into further division and intensification of design work, in the long run, proves to be inferior to a strategy which increases the competence of self-organisation and innovation, emphasising the special human abilities of learning from errors or deviations and shaping new ways of productive and qualitative work procedures[2].

But what do we mean by *work-culture based experience and action of designers* ? And how may we be able to develop a methodology which enables the researcher to be in continuous dialogue with practical experienced industrial designers through all the R&D phases? Basically, the Anthropocentric approach implies an understanding of the industrial design process and the industrial designer as elements of a *work culture*. This means that the designer's beliefs, knowledge, actions and tools are partly based on traditionally accepted norms and practices, but are also subject to changes demanded by new external conditions. Thus, the concept of work culture includes both material aspects (e.g. tools used by the designer, products designed) and mental aspects (e.g. how the designers practise design, how they may learn new kind of practice).

Methodologically, the Anthropocentric Approach departs from the 'objective' paradigm of conventional automation approaches. Designers are not perceived to be passive objects, which are acted upon, but rather as subjects able to act and participate in the process of dialogue. Neither can the researcher remain aloof as a 'neutral observer', but must also participate in the shaping process.

The Anthropocentric Approach implies transcendence of traditionally bound barriers between theoretically and practically experienced knowledge. Thus new methods of dialogue are needed. Research becomes not only a question of analysing and evaluating but also of shaping new visions of work organisation and technical systems/tools in a continuous dialogue with practically experienced people involved.

The following case of how the concept and demonstration model of an electronic sketchpad was developed may serve as an example of how the Anthropocentric Approach can be elaborated in practice[3].

The concept of the Electronic Sketchpad Pad

Emergence of the Idea

Initially, the approach was based on the negative experience of how CAD systems limit the creative activities of industrial designers in certain phases of the design process,. As mentioned earlier, CAD systems have a tendency to promote isolation, induce passivity and to alienate the designer from the world of machines and equipment and their operational details. Although CAD systems improved the speed and ease of modifying and plotting a drawing and increased the possibilities for combining two drawings and/or re-scaling a drawing, it proved to be an inadequate tool in the more creative phases of design.

Moreover, compared with the possibilities of an overall view at the traditional drawing board, the screen was experienced like a microscope. Many designers expressed as negative response of the so-called 'key-hole' effect, that is, they felt they were peering through a keyhole when working with the CAD system.

In order to shape a more human centred piece of equipment, the following idea was advanced[4].

The idea was to combine the positive aspects of the traditional drawing board and of the CAD-system. This offered the designer the possibility of using the drawing board as a digitizer during the design process, while employing the CAD-system for the preparation and refinements of the final drawing.

This initial idea can be briefly described as follows: When the basic geometry of a part has been determined, it is drawn at an electronic drawing board and simultaneously digitized. The digitized geometry is transferred to a CAD-system where changes and annotations can be made. The drawing board/digitizer acts as a pan-plotter as well, which makes it possible to retrieve (standard) parts from the CAD-system and add/insert them into the drawing. Finally the finished annotated drawing can be plotted automatically by the drawing board.

The original idea, as presented above, was developed experimentally in various ways during the first year of the project[5]. Six different possible solutions to the problem were worked out. Of these, the solution presented above was selected as the most viable at that time. The selection was made by the technical groups alone, based on their expertise, on existing hardware and software and on development tendencies.

Involvement of User Groups

While the concept of the electronic drawing board was being developed technically, groups of users from different Danish undertakings were established by the social science group. One group consisted of Danish technical school-teachers, experienced in teaching the use of CAD-systems. Two other groups consisted of technicians engaged in different types of alternative design and drawing. These groups were invited to evaluate and otherwise comment upon the technical approach in February, 1987.

The first reactions were not positive, partly due to misunderstandings. Members of the user groups, especially draftsman working with final drawing/documentation expected a new, complete CAD-system implemented on the drawing board while, in fact, the intention of the electronic drawing board was to remove the 'key-hole' effect of existing CAD-systems, but not to try to compete with their effectiveness in dimensioning, hatching, editing etc.

The research group arranged a new meeting in March, 1987, with only one member from each of the three user groups in order to explore the problem in depth. At this meeting, the representative from the designer group suggested improving tools of communication between designers and the pattern makers. He wanted a handy, portable electronic sketch pad to facilitate the personal communication he was used to having with colleagues and, particularly pattern makers. Furthermore, he expressed interest in testing the idea in a practical context.

The idea of an electronic sketch pad, handy to carry and easy to use, emerged from these discussions.

The Laboratory Model of the Electronic Sketch Pad

The main purpose of the electronic sketch pad is to record the initial sketches made in the design process and during discussions with process planners, pattern makers and NC-programmers/machinists.

In this way, the electronic sketch pad becomes a tool for the designer/developer and not for the draftsman. Recorded sketches can then be transferred to the CAD-system for further detailing and annotation.

The intention of this development was to create a new system -the electronic sketch pad- in line with the following objectives:

1) The system should be simple to operate and easy to learn.

2) The system should resemble the traditional sketch pad (pencil and paper).

3) The sketches made should be recorded and stored in a compact, computer readable form.

4) The system should enable previous sketches to be retrieved and displayed for further elaboration.

6) It should enable retrieval of the geometry of (standard) parts from CAD/CAM-systems for insertion into a sketch.

7) The system should have display facilities.

8) The electronic sketch pad should be portable.

9)The stylus should resemble a normal pencil, i.e. with no cables attached.

10) It should enable the exchange/distribution of sketches between designers and process planners etc.

It was not among the objectives of this development to create and implement yet another CAD-system (D for drafting). Rather, the development of the electronic sketch pad should be regarded as a step in the development of flexible user-interfaces and flexible input media for a new generation of design systems. The principle of this system functionality is illustrated in enclosed fig.1.

As illustrated in fig.1 the proposed sketch pad makes it possible to use sketches, previously executed drawings or IGES-translated CAD-drawing as inputs from which the design of a new component can be made. The geometry should be drawn up and edited either by the designer or by the work preparer, and be mutually exchanged. Because the sketch pad system is the same for both parties they can easily communicate and alter the design by means of a common language. The sketch pad is easy to carry around if so required, and thus easily allowing informal personal communication in contrast with more stationary CAD-systems.

Furthermore, the intended sketch pad may enhance learning possibilities by adding at least two dimensions. Since learning its use is relatively simple, it may be advantageous to introduce the system to designers who are not familiar with the CAD-system so they can gradually learn the computer's technical facilities. On the other hand, those well versed in CAD-systems but unfamiliar with the work culture of the design process may be gradually introduced to the latter by employing the sketch pad with the help of a more experienced designer.

At the outset of the design process, the designer is expected to utilize sketches to express his ideas. At times, he or she may prefer a traditional sketch pad before the electronic sketch is used. At other times the sketch pad may be used from the outset.

Fig. 2. Demonstration model of the electronic sketch pad.

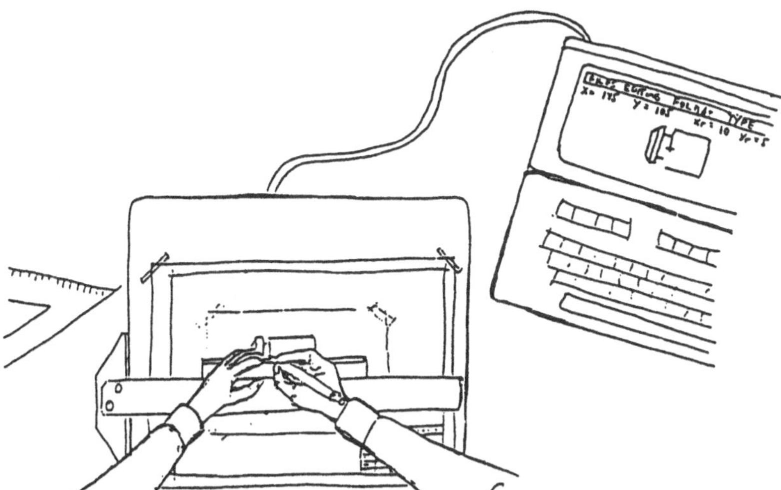

These ideas can be drawn up in shapes of various items on the sketch pad and stored in the computer's 'idea bank'. The idea bank can be distributed to other members of the design team for further work. After deciding upon the shape of the part it is drawn in its exact geometric shape on the sketch pad. At this time, the drawing can be transferred to the CAM-area to be examined and modified. During this process the designer and the production planner can work with the shape and use the designer's notes on tolerance, surface finish, etc. in order to obtain a part which can be expected to be produced by the machinery. The drawing is then transferred to a CAD-system for additional information to complete the documentation of the drawing.

Further Improvements to the Electronic Sketch Pad

Economically speaking the electronic sketch is a small, flexible and relatively inexpensive instrument. Potentially, it can also reduce the comparative amount of time used on the larger, costlier CAD systems. Thus, it can contribute to a more economical use of time and resources in the drawing office. So far, the electronic sketch pad is developed as a demonstration model. Further development may include improvements of the following kind:

Higher sample rate from 50 to over 200 per sec. during the digitilizing of movements may improve the sketch pad's ability to follow the movements of the hand.

Higher resolution on the display (from 0.3 * 0.35 mm pixel to 0.1 * 0.1 mm pixel size) may improve the quality of the picture to the limit of the eye.

Better editing facilities like direct erasing possibility with the cordless pencil/eraser, and an interactive editor may improve the spontaneous mental activities during the creative phase of design.

A more stable *pattern recognition*, better *snap routines* and *scale* of different sizes in the same sketch are necessary according to the usefulness of the tool.

A *database* for filing/query on sketches (e.g. idea, personal schedules, memo) may facilitate the opportunities of storing and retrieving sketches as the designer may prefer in a given situation.

Recognition of handwritten text and update IGES (Initial Graphics Exchange Specification) for *text exchange* may create the possibility of handwritten notes on the drawing.

Direct communication for file transfer to CAD systems on a single command and a *handheld battery operated* device with combined display/digitizer with a cordless pencil, and up to electric or wireless connection to a computer network node make it possible to use the electronic sketch pad as a portable tool, in situations where the designer wants to discuss sketches with colleagues in other departments.

The electronic sketch pad may be viewed as a modest example for improving the technical facilities supporting creative and communicative design activities. The realisation of this aim depends not only on its technical functionality, but on the social worklife and work culture established in designing departments.

Anthropocentric Principles of Worklife and Organisation

The Concept of Worklife

Worklife is a concept with many connotations. Basically, it includes both explicit, regulated practices as well as non-formalized, situational and spontaneous activities and expectations. Far from all aspects of worklife are visible or even consciously recognized. As a complex of individual and collective actions and social relations worklife is an important part of workculture.

During the prototyping of the Electronic Sketch Pad a number of issues about social worklife in the Drawing Office came to our attention. Would it be possible to synthesize these issues on a common vision based on anthropocentric principles of organisation? We decided to try[6].

Organisation of Product Development

According to the anthropocentric orientation emphasizing the special creative and communicative abilities of industrial designers, product development should be organized in semi-autonomous project groups consisting of representatives from all relevant departments right from the outset of the development process[7].

Instead of a fixed division of tasks, a multi-functional team collectively possess skills and abilities needed in development of a product in a pattern of *overlapping* skills and knowledge. Moreover, this kind of organisation may *transcend* traditional, hierarchical divisions between development, work preparation and production and thus create a more flexible structure for a direct communication network. The project group should assume responsibility for all decisions regarding detailed time scheduling, expenditure and purchase of relevant tools, materials and technical equipment.

Balance Between Individual and Collective Considerations

Group autonomy is a central concept in anthropocentric oriented visions. But this principle is not unambiguous. The designers' wish for close cooperation was not the only one. A potentially opposite wish for extensive individual autonomy was also articulated. It may be fine to work in groups, but there must be time and space for working alone and undisturbed too. The detailed researcher-designer dialogue during the vision workshops revealed how important the subjective factor is regarding concrete problem solution. The individual differences in work method, social attitude and ability to cooperate should not be reduced to 'disturbing factors'. As far as possible, the individual designer should be able to choose his or her own method of work. Particular abilities should be respected and encouraged in the frame of cooperative endeavour. The encouragement may take place through a continuous process of learning and renewal in the department. By organising leeway for experience, problem-solving and imagination, the working methods of the department can be continuously revised and developed.

Though video-telephone and electronic mail-system may increase opportunities for quick communication, the importance of direct face-to-face communication between members of the project group was heavily emphasized by the industrial designers. Unless actively counteracted the use of CAD may discourage face-to-face communication and thus in inhibit rather than enhance innovation and flexibility. The electronic sketch pad is an example of counteraction in this respect. But more fundamentally, organisational initiatives must be included too.

An anthropocentric oriented workprocess presupposes vocational training and education in a close and dynamic interchange with the daily work situation and subjectively bound skill improvement. How can the conditions for such an interchange be provided? During the workshops conducted between researchers and industrial designers the following suggestions were advanced:

- *every individual in a department should have the opportunity to define his or her need for new skills and/or knowledge. Such requirements can be forwarded to the department of vocational training in abbreviated form.*

- *the department of vocational training could then develop a series of courses, together with the individuals who made the initial requests.*

- *collective departmental participation in the same course (for example in communication and project team-work) could facilitate systematic implementation of the content of the courses afterwards in daily work.*

- *project teams could be structured in such a way that practical experience 'from inside' and new knowledge 'from outside' would be exchanged between 'old' and 'new' employees of the company.*

- *special facilities for experimenting with new systems, tools and methods before they are implemented in the daily work situation should be provided.*

The anthropocentric vision encourages a 'bottom-up' or participation approach, where ideas, practical experience and theoretical knowledge may emerge from ongoing organisational processes. Instead of imposing predetermined means and goals, it may be far more productive to devise means where subjectively-bound practical experience and more-or-less formalized knowledge can be combined in a reflective and often innovative spirit of cooperation.

Methodological Challenges

User-Researcher Cooperation

The Anthropocentric Approach implies, as previously mentioned, transcendence of tradionally bound barriers between theoretically and practically experienced knowledge. Thus, user-researcher cooperation is inevitable. It is impossible to design future work places for human needs without openly discussing the issue with those possessing the practical experiences.

The *initial clarification* of the problem of the research project is a far more open and sensitive phase in the anthropocentric approach than in traditional, positivistic approach. The *objective* is only defined in general terms in advance. Different intersets and resources of researchers and users in cooperating often need to be discussed and modified according to commonly defined objectives and processes of dialogue.

Different models for cooperation may be prioritised during different conditions: A *minimal* model where the researcher alone takes it upon himself to interpret the results *for* the user. A *maximal* model where the researchers cooperate with the users in all the phases of the project. Or a *combined* model where the users participate in parts of the project[8]. In principle a maximal model may provide the most appealing opportunities of transcendence and innovation. Often a combined model may be chosen for pragmatic reasons (shortage of time and resources).

Limitations and Promises of the Dialogue

During the dialogue process several methods may be used (e.g. experimental workshops, prototyping, organisational play, multi-metaphoric approaches[9]). The common feature of such methods is the open, experimental and interactive character. If the positivistic criteria of intra- and intersubjectivity are applied, the dialogue method will rank rather low on reliability. This should not be a source of worry, however. As Galtung explains:

> *... The dialogue is dynamic, flexible and self-transcending, and should be so - hence these methodological criteria - in a sense - do not apply. The task is to produce, not to reproduce*[10].

Compared with the traditional survey method, the dialogue possesses opportunities for creating deeper, more reflective and more action oriented insights and/or models of innovative tools and organisational structures. By introducing theoretical theses and participating in the user discussions of the issue from a practical viewpoint, the researcher does not only get an answer to his prefabricated analytical categories but also an understanding of the actual complexity of the concrete case in question.

The scientific rationalistic aim for order and analytical understanding are continuously reminded of the value aspects: no one 'right solution' exists. But ideas based on different interests and values may be confronted in a dialectical process and thus transcend the limitations of one single perspective and move towards a more promising synthesis on a higher level of understanding and action.

Conclusions

One may ask whether the anthropocentric approach will penetrate the dominating scientific and industrial traditions of mechanistic, technocentric perspectives and practices? For centuries the Newtonian-Cartesian paradigm has been so powerful in its pragmatic technological applications that all research challenging this dominant paradigm has been suppressed or encapsulated as minor modifications. However, the question is not only of theoretical interest. More and more the technocentric approach seems to create various kinds of obstructions mitigating against its own generally defined aims of 'predictability' and 'efficiency'. Some industrial firms and managers have realized the need for:-

- *'less hierarchical and less compartimentalized organisations for greater flexibility'*

- *'.. human resource policies that promote continuous learning, teamwork, participation and flexibility'*[11]

Statements like the above mentioned are not in itself a guarantee of fundamental change of strategy and paradigm towards a more anthropocentric orientation but they may point towards, at least, greater openness and better opportunities of experimenting with new kinds of methods, tools, organisations and educational approaches.

References

Brödner, P (1986) *Fabrik 2000* Wissenscaftszentrum, Berlin

Cooley, M (1981) The impact of Computer Aided Design on Designers and the Design Process. Unpubl. Ph.D., North East London Polytechnic, London

Cooley, M (1987) *Architect or Bee?* The Hogarth Press, London

Dertouzos et al (1989) *Made in America* MIT commission on Industrial Productivity. Cambridge, Massachusetts, The MIT Press

Ehn, P (1988) *Work-oriented Design of Computer Artefacts* Arbetslivscentrum, Stockholm

Eriksen, E R (1982) *The drawing office of the future* Technological Institute, Tåstrup, Denmark

Finne, H (1982) *Mellem tegnebrett og terminal* IFIM, Trondheim, Norway

Floyd, C (1984) *A systematic Look at Prototying* Spinger Verlag

Galtung, J (1988) *Metholdology and development vol. III* Christian Ejlers, Copenhagen

Læssøe, J & Rasmussen, L B (1989) *Human-centred Methods* IS, DTH, Copenhagen

Læssøe, J, Rasmussen, L B & Tøttrup, P (1989) *The Electronic sketch pad* IS, DTH, Copenhagen

Majchrzak, A (1987) *Human aspects of Computer-Aided Design* Taylor & Francis, Philadelphia and London

Morgan, G (1986) *Images of organisation* Sage, London

Noble, D.F (1977) *America by Design* Alfred A.Knopf, New York

Noble, D.F (1984) *Forces of Production - A social history of Industrial Automation* Alfred A.Knopf, New York

Rasmussen, L B et al. (1987) *Work Culture and CAD requirements* ESPRIT PROJECT 1217 (1199) Deliverable R12, DTH, Copenhagen

Rauner, F, Rasmussen, L B, Corbett, M (1988) The social shaping work and Technology. In *AI & Society,* Febr.1988.

Volmerg, B, Senghaas-Knoblach, E & Leithäuser, T (1986) *Betriebliche Lebenswelt.* Westdeutscher Verlag, Opladen

Wingert, B and Riehm, U (1985) Computer als Werkzeug-Anmerkungen zu einer verbreiteten Missverstandnis. In Rammert, W.et al. (eds.): *Technick und Gesellschaft.* Jahrbuch 3, Frankfurt

Weizenbaum, J (1976) *Computer Power and Human Reason - from Judgement to Calculation* W.H. Freeman and company, San Francisco

Winograd, T, and Flores, F (1986) *Understanding Computers and Cognition - a New Foundation for Design* Ablex, Norwood

Notes

[1] The problems are not created by the CAD systems of course, but are part of a more long term development dominated by the mechanistic or technocentric model of rationalisation.

A number of studies since the 1950's were done in this way attempting to simplify drawing and design methods. Since the mid-seventies, Tayloristic principles have been applied methodically to analyse and formalize activities in the design process. CAD systems can be seen as a further step in this direction.

[2] This is not only a phantasy of idealistic academics. In the MIT commission on Industrial Productivity (1989) *Made in America*, the central conclusion is that the technocentric approach has failed in many parts of American industry. The report recommends a strategy much more focused ion the human resources.

[3] The electronic sketchpad is developed by a Danish research group as part of the ESPRIT project 1217 (1199) *Human-Centred CIM systems* during the period 1986-1989. The project was directed by the GLE, London as prime contractor and Mike Cooley as director. German and English research groups and industrial companies participated as well with regards to CAP and CAM systems. The Danish group consisted of social scientists and engineers from the Institute of Social Sciences at the Technical University of Denmark, engineers from the Technological Institute and NEH Consultant Engineers and industrial designers from LD Ltd. The project was financially supported by the the Technology Council in Denmark and the EEC Commission.

[4] This idea was first presented by Erik Rask Eriksen in *The Drawing Office of the Future* (1983, TI, DK).

[5] Some of these developments are described in detail in the Technical appendix to R2 "Human-centred input media for CAD systems", Copenhagen, 1986 (unpublished)

[6] The following visions are based on seven meetings or 'vision workshops' conducted by two worklife-researchers, one system developer and two industrial designers. The results of this process is described in further details in Jeppe Læssøe, Lauge Baungaard Rasmussen, Poul Tøttrup (1988) *The electronic sketch pad* p.81 ff. ESPRIT PROJECT, Deliverable 19, Copenhagen.

[7] An example of the make-up of such a multi-functional team consists of representatives from the development department, plastic department, metal department, purchasing, dispositions, planning, quality control department, sales department, production planning and production

[8] For a more detailed discussion, see Jeppe Læssøe, Lauge Baungaard Rasmussen (1989) *Human-Centred Methods* p.41 f.

[9] These methods are explained in Jeppe Læssøe, Lauge Baungaard Rasmussen op.cit.

[10] Johan Galtung (1988) *Methodology and Development* Vol. III, Copenhagen p.82.

[11] MIT Commission on Industrial Productivity (1989) *Made in America* p.118 f.

Fig. 1: Flow process of design work and work proparation for a new design.

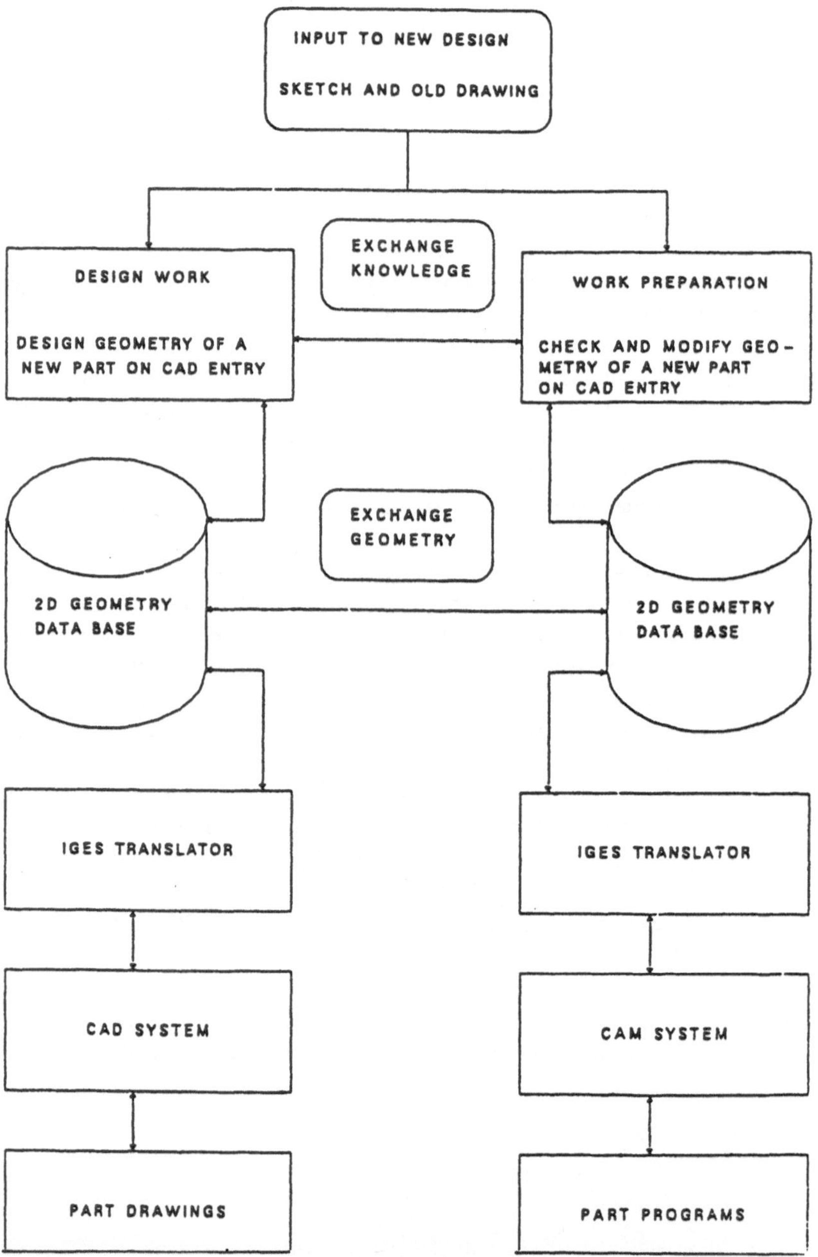

Skill in Software Production - The Deskiller is Deskilled

Yoshihiro Sato

Introduction

Programmers participate in the larger part of the labour of software production. Since the data processing technology has advanced at an exceptional speed, manual programming cannot accommodate the demand for software: In Japan, it is estimated that shortfall for approximately 970,000 software engineers will occur by 2000. The 'Nikkei Computer' magazine reported that the backlogs accumulated in companies was 550 person-months/company on an average. Affected by the shortage of software engineers, increases in the number of engineers, and development of productivity improving technology are attracting attentions in the field of software production.

Because of the software crisis, various measures have been taken to improve productivity. As a result, then use of languages for end users and application of software development supporting tools has accomplished improvement of the productivity and quality and reduction in the delivery period and have finally resulted in reduction of the workload. However, reduction of the work did not apply to programmers but increased the absolute job volume. These problems of programming seriously affect the quality as well as the quantity of labour. Although technology development in the programming process enhances the program producing power, the level of skill required for programming has been lowered at the same time. Changes that occurred in other fields such as manufacturing using computers are also occurring in the programming field. Historically, computers divided labour in various fields and deskilled the workers. Also, 'the deskiller is deskilled.'[1]

Programming and Skill

ENIAC exhibited the calculation process with complete formula in 1946; it had 6000 switches. Instructions for the calculation procedure were enacted by changing over the switches and power distribution board. On ENIAC, the connected circuit itself was the calculation procedure. Instructions for the procedure (i.e. programming) were enacted by the person who understood the computer including hardware. Designing the calculation procedure was programming and was a visualized work.

The idea of the 'program built-in system' established by Von Neumann et al. changed the form of instructions for calculation procedure from switches and wiring to a program.

In 1944, the world first program built-in type computer, EDSAC, came out and the prototype of the assembler language was developed. Assembler languages at an earlier stage are Soap for IBM650 in 1954 and Autocoder for IBM705 in 1955. The first practical assembler language is said to be SAP (Symbolic Assembly Program) for IBM704 in 1956[2].Since the assembler language is a programming language that is at the closest position to machine language, an instruction of the assembler language basically corresponds to an instruction of an machine language. The machine language is expressed by binary digits, while the assembler language has been coded by alphameric characters. The change from binary digits using O and 1 to coding using alphameric characters is basically a matter of notation and it can be considered on an extension of the initiatives by Pascal and Leibniz. When assembler language is used, the programmer is required to perform programming while recognizing addresses and registers because instruction units of the assembler language are basically same as those of the machine language. It means that programming is greatly dominated by the programmer's experience and knowledge.

Assume a program for calculation of 'A = B + C; D = A' as a very simplified example. If this process is coded by using the assembler language, the instruction sequence is as follows:

(l) Calculate 'B + C' on a register
(2) Store the register value in the memory address that corresponds to variable A.
(3) Load the value of variable A in the register.
(4) Store the value of A in the address that corresponds to variable D.

This is a basic form, so the beginner complies with it. However, a skilled programmer recognizes program optimization and would consider omitting the process in (3). After the process in (1), the calculation result of 'B + C' is retained in the register, then processes (2) and (4) are done. As a result, a loading operation is not necessary. Reduction of loading operations is very significant. Particularly, optimization on the working set of the on-line program increases the processing capability. To apply this technique, the intermediate data on the register should be retained until the processing time. However, the number of registers available for the programmer is limited. Therefore, the programmer cannot retain much intermediate data unlimitedly or delay the processing timing unlimitedly. The skill of optimum register assignment based on the skilled programmer's experience and knowledge is required. Other optimization techniques are factoring of the common part, removing the invariant away from the loop control range, change from multiplication and division to addition and deduction, storage area dynamic allocation, etc. Each technique requires decision making that takes situational factors into account and the skill is supported by experience.

Building an effective logic through combining basic instructions is done by knowledge backed up by experience. A program that is free from syntax errors and satisfies the specifications will work without trouble no matter how programming was done. However, program quality greatly depended on the capability of an individual based on his/her experience. Once a program in a form of a file is stored in the computer, it is hard to check the quality because the information is in magnetic form. But the quality can be easily known from the operation efficiency, ease of maintenance, and function expandability. Programs at that time were handicraft based on creativeness and imagination, like masterpieces of artisans. A program configured with an error-free excellent algorithm was applauded with the word, 'masterpiece'. In that sense, the programmer who made a program by using assembler language was a technician having expertise. Then, what is essential in programming using the assembler language?

It is logical understanding of the computer, that is understanding of the operating system (OS). A program operates under the OS that is the software environment which surrounds the program. Since a program developed by using the assembler language is very close to the machine language, the program operates while closely communicating with the OS. The supervisor may be often called directly by using the SVC macro instruction. File open/close is one of the typical processes. in this case, the program must prepare the control block as the interface with the OS[3]. See enclosed Table 1.

The programmer should also know the meanings of fields in the control block and the single-bit flags. Either O or 1 must have been set for each flag. File open may end abnormally even though a single-bit flag has been reversed. Since flag meanings can be understood based on OS knowledge, the programmer must systematically know not only the OS as the core of computer operation but also the grammar of the language.

Programming using assembler language cannot be learned in a day. Before execution of a job, an approach run that is learning and training is indispensable. A programmer understands the computer as a system by learning and hands-on training, and at the same time, acquires the expertise on programming techniques.

Productivity Improvement By the Language Development

The first breakthrough in programming languages was FORTRAN-I developed for IBM704 in 1957. FORTRAN is a typical language for scientific and technical calculations that has eliminated the register and address operations. The impact of FORTRAN can be understood from the fact that market evaluation of UNIBAC and IBM changed after the introduction of this language[4]. After a while, a high level language appeared for office work. The CODASYL (COnference on DAta SYstems Language) as an organization to develop the common language for office work disclosed COBOL (COmmon Business Oriented Language) to public in 1960. Then, the standard has been changed several times. Presently, COBOL is most popular as the office work program development language. It is said that 90% of programmers in the world are using COBOL.

Since instructions can be described in English in COBOL, productivility was improved to twice as much as that of the assembler language[5]. However, addresses and registers are not controlled by the programmer in FORTRAN and COBOL, the field of the program optimization technique to show the programmer's skill was narrowed. For example, the process, 'A = B + C; D = A', brings the same logic no matter who codes it and there is almost no space for optimization. If several programmers encode a 200- to 300-step program, some programmers may make the same program.

The OS knowledge level required by programming has been extremely lowered. A high level language is, so to speak, a preprocessor to create the assembler program. Since use of the supervisor call and preparation of the control block are supported by the preprocessor, programs need not recognize the OS.

Software Development Supporting Tool

As in other fields such as product manufacturing, automation by computers has been introduced in the field of software production. Software to automate software production is called *software development supporting tools*. These tools are used in the processes such as request definition, basic design, detail design, programming, test, and management that configure software production. Among them, the tool that supports the programming process enables the programmer to execute the job with a low skill level.

Also, the preprocessor that automatically creates a program is one of the software development supporting tools that assist the programming process. Compared with high level languages such as COBOL, the preprocessor is called the super high level language. Typical super high level languages in Japan are HYPERCOBOL of Fujitsu and IDL of NEC. Since the typical processing logic of the program has been put into a macro, the volume of coding is half as much as that of COBOL. The deadlock exit routine is provided in the program even though the programmer has no knowledge about deadlock.

There are software development supporting tools that have eliminated programming itself. COMPAL of Fuji Software automatically creates a program in C language from the program chart created in the interactive mode on the terminal. YPS of Fujitsu creates the COBOL source from the program chart created on the screen. Mental labour in programming seems to be decreased more and more. See Figure 1:

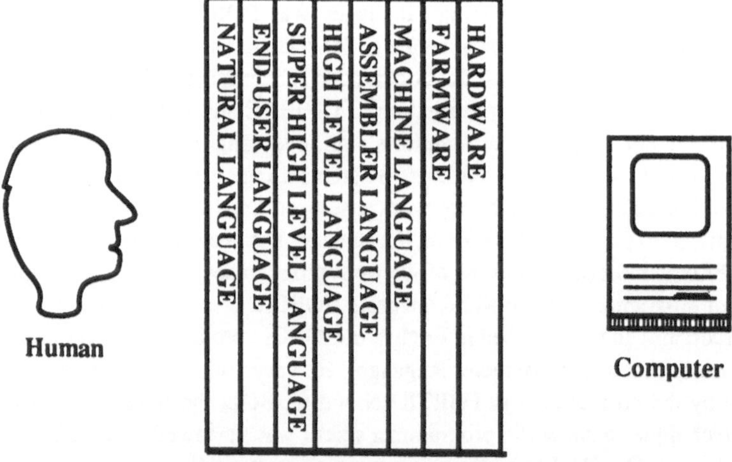

Figure 1. Programming Language

Deskilling

In the past, programming was the operation to stack instructions as minimum elements while combining them according to expertise and experience. Presently, programming is the operation to arrange the blocked process units orderly and efficiently. Although a pattern can be created in such a way as handling the LEGO SET, the mode of properties of the element itself cannot be changed. Further, software development supporting tools that have eliminated programming itself are available. There is no significant difference between a program made by a beginner and a program made by a skilled programmer. However, it does not mean that the skill of the beginner is excellent. It means that the level of the skill required for programming has been relatively lowered in software production. An aggregate of situational factors and decision making is provided as a unit in advance. The programmer is required to select and rearrange the unit adequately. In the field of programming where amateurs are increasing, the quantity and quality of the skill required are changing.

What is the Programmer's Skill?

It has been stated that measures for improving productivity deskill programmers. What is the programmer's skill in software production?

The programmer's job is to produce a program according to the specifications. The systems engineer (SE) who designs the system determines the program specifications. According to the specifications, the programmer creates a program. In that sense, the programmer is similar to a latheman who produces a product according to the specifications. The programmer takes actual actions such as decision making on programming based on the expertise about the job. 'Expertise' referred to herein is what can be clearly described as rules. It is the explicit knowledge that can be converted into the knowledge base.

The tacit knowledge resides in between the explicit knowledge and actual actions and connects them. The skill consists of the expertise based on common sense, tacit knowledge formed through experience, and actions as a result. See Figure 2:

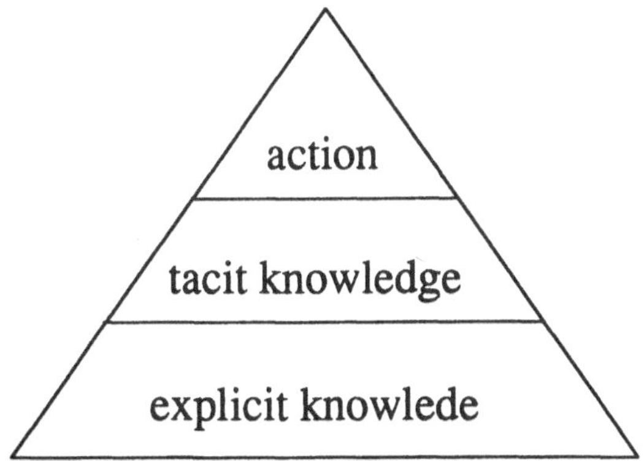

Figure 2. the programmer's skill

The expertise that is the explicit knowledge for the programmer is what is essential in the job. The expertise is shown in manuals such as the programming language reference manual, program specification, debugging method, TSS terminal operation, etc. Actual actions based on the expertise such as coding, testing, debugging, etc. are taken for various problems. Through gaining experience and repeating training, the characteristics of the connection between the expertise and actions vary. The programmer gradually becomes capable of taking optimum and effective actions according to judgment corresponding to the situation. This is the process of forming the tacit knowledge and can be called 'learning' or 'training'.

The beginner thinks how his accumulated knowledge can accommodate. If the volume of learned expertise (rules) is small and the tacit knowledge that would lead judgment has not been formed, the beginner can recognize and judge what matches the given conditions but cannot recognize and judge other matters properly. Feedback of the optimum answer for a new situation changes the characteristics of the connection between knowledge and judgment. This cybernetic loop is the 'experience'.

Skill Change: Deskill or Advance?

Improvement of software productivity has changed the programmer's skill. The change from the assembler language to the high level language enabled programming even though the programmer has almost no knowledge of programming techniques such as optimization and the OS. By using software development supporting tools, the programmer can accomplish programming even though he/she hardly knows the grammar on the programming language. A beginner can make a program that is equivalent to what is made by a skilled programmer. It means that the beginner's skill is not high but the programmer's skill level required in software production has been lowered.

The programming operation is increasingly replaced by machines. Writing in the coding sheet was replaced by entry to the TSS terminal. Logic examination was changed to examination of application to tools. In software production, terminal operation and tool utilization are becoming more important for the programmer that the OS knowledge and programming techniques. The change of labour in the programmer field seems to make the experience and expertise unnecessary.

In considering the programmer's labour, the viewpoint on the change of means only does not clarify human being. What must be checked is the human's skill. The skill is deeply associated with human's labour. Technological innovation results in a new means. The change of capability difference between the beginner and skilled person is important in the workshop which uses a new means. 'The new art and tacit knowledge would arise to accompany and allow to use the increased explicit knowledge".[6]

The problem is not increase or decrease in the tacit knowledge but its quality change. Checking the quality of the tacit knowledge may bring ranking of jobs and prejudice. However, a discussion insisting that any job has tacit knowledge does not clarify the effects of the technology on human labour. I dare say the quality change while highlighting the skill level. If a beginner can do the same job as a skilled person because of the changed means, the skill level for the job is lowered. In the workshop where jobs can be accomplished in the beginner's skill level, can people with full experience continue the operation? As the background of 'programmer's retirement at the age of 35', not only the severe working conditions for 'workers on desk' but also the present programming situation that eliminates improvement of the skill should be considered as important aspects.

Skill Transfer

Expertise of skilled programmers have been implanted into computers in the form of software development supporting tools. Of course, implanted expertise is the rules using limited situational factors as parameters. The process of determining each weight of unlimited situational factors has been overlooked even though the knowledge base covers all of the important situational factors as the decision factors. Really, this process is important. The quality of the skilled programmer's expertise is obliged to change when it is converted into a program. The skilled programmer's decision making process is different from the logic of software development supporting tools.

A skilled programmer can perform virtual programming according to the required specifications. As a means to embody the concept, the skilled programmer uses the tools. If a beginner uses the tools, the same quality software may be produced as a result. However, the process is completely different. The beginner obtains the knowledge about entry of input data and fetching of output data and the knowledge about tool operation, but he/she cannot obtain the knowledge in the process that assembles a program from the specifications. (See Figure 3.)

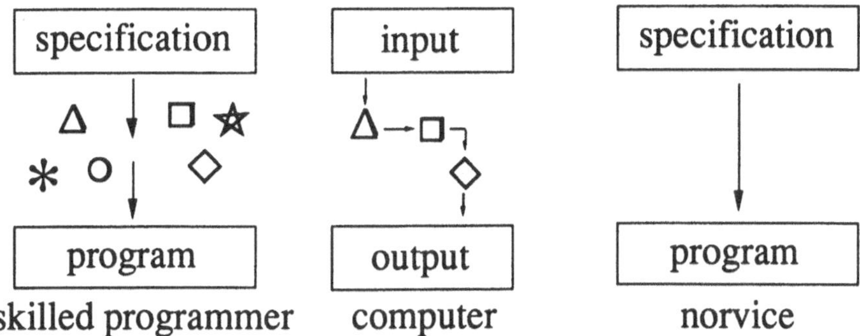

Figure 3. software development supporting tool and skill

Skill Improving Labour

Because of the chronic manpower shortage in the information service industry, tools have been developed so that a beginner can do programming after short-time education even though he/she has little knowledge and poor experience. Computers have accumulated the skilled people's expertise in a form as programs. When a job is done by using a computer, operation takes precedence in the job. In the programming field, computer installation has made the job greatly dependent upon the operation like in other fields. Same as workers in other fields, programmers who work using a computer are required to obtain the knowledge about operating various tools rather than the knowledge of the job. Programmers can execute jobs without expertise. Simplification of programming is good news to end users who try to develop programs by themselves. Then, how does it affect to programmers who are engaged with programming? Lowering of the skill level required for execution of the job makes the skill like a fossil and brings amateurism in programming. Creativeness and imagination are important factors in human labour. Labour which has less opportunities and extent for decision making deprives creativeness and imagination from human being. Regret to say, the super high level language that arranges units or the present tool using chart input is the work that fits the specifications to input rules. Compared with programming using the assembler language which creates various techniques from experience based on expertise and makes programs according to the programmer's skill, there are few decision making parameters valuable for present programmers.

In software production, when 'hand' and 'brain' were separated from each other, deskilling of programmers began. As tools (languages and tools) are improved and productivity is improved, programmers' creativeness and imagination became unnecessary gradually. Participation by the human being and computer is very important in this aspect.

Programmer's Decision Making

Symbiosis of man and machine must be considered also in software production. It should be examined not only from the viewpoint of the interface between the man and computer but also from the viewpoint of the entire software production labour. Although positioning, roles, and functions of the computer are the factors that should be considered in the anthropocentric context, the solution lead from the computer interface in labour as a part of the process could be a first aid but would not eliminate the problems. Separating a particular process from software production and applying the anthropocentric idea to programming only would be impossible. Because the anthropocentric idea does not correspond to each divided operation. In other words, dividing the labour itself is being objected to. In software production, catching the users' needs through operation and evaluation should be considered as the system life cycle. (See Figure 4)

L	PN			DN						PG	TG			OT	ME
M	SP	PP		SD		SRD	PD		PDR	PG	IT	ST	TGR	OT	ME
S		BP	DP	ID	LD		PS	MD							

PN:Planning
SP:Survey and Planning
PP:Project Planning
BP:Basicproject Panning
DP:Detailproject Planning
DN:Design
SD:System Design
ID:Initial Design
LD:Logical Design
SDR:System Design Review
PD:Program Design
PS:Program Structure design
MD:Module Design
PDR:Program Design Review
PG:ProgramminG
TG:TestinG
IT:Integration Test
ST:System Test
TGR:TestinG Review
OT:Operational Test
ME:Maintenance and Evaluation

Figure 4. the process of software production

As explained, programming is an important process in software production but it is a part of the entire processes. The change of the means to sharpen a pencil from a knife to an electric sharpener is merely the changes of the means in the sharpening process among the entire processes. The reason the languages or tools are important in software development is that programming is not a part but the whole of the process for programmers. The changes of the means cannot be avoided when technologies advance. The reason the improvement of the means not only results in changes in the process but also lowers man's working skill is that labour has been divided.

Introduction of Designing Divided Labour into Mental Labour and Muscular Labour - A Conclusion

Software production should be reunified so that programmers will be associated with labour and have responsibility through the life cycle. Labour that has been horizontally or vertically divided into segments becomes significant when the segments are coupled from each other. Treating each individual segment is inadequate. Not to deskill the deskiller, it is necessary to see workers not as a quantitative resource such as productivity but as a qualitative resource having the tacit knowledge inherent to human being and knowledge based on experience. The technological vector must be determined so that programmers can participate in *all* processes in software production.

Offset / Field				Section
0(0)	FCBRELAD			Device Dependant Section (Directry Access Storage Device Interface)
4(4) FABKEY	5(5) FCBFDAD			
	13(C) FCBDFCTAD			
16(10) FCBKEYLE	17(11) FCBDEVT			
20(14) FCBBUFCB FCBBUFNO	FCBBUFCA			Access Method Common Interface
24(18) FCBBUFL				
28(1C) FCBJOBAA FCBLNP	FCBIOBAD			
32(20) FCBEODAD FCBFALN	FCBEODA			Foundation Block Extention
36(24) FCBEXLST FCBRECFM	FCBEXLSA			
				Foundation Block (after OPEN)
44(2C) FCBDEBAD FCBIFLGS				
48(30) FCBREAD/FCBWRITE/FCBGET/FCBPUT FCBOFLG	FCBREADA/FCBWRITA/FCBGETA/FCBPUTA			Access Method Dependant Section (BASM, BPAM, QASM INTERFACE)
52(34) FCBGERR/FCBPERR/FCBCHECH FCBOPTCD	FCBGERRA/FCBPERRA/FCBCHECHA			
56(38) FCBSYNAD FCBIOBL				
60(3C) FCBCIND1	61(3D) FCBCIND2	62(3E) FCBBLKSI		
64(40) FCBWCPO	65(41) FCBWCPL	66(42) FCBOFFSR		
68(44) FCBIOBA FCBIOBBT	FCBIOBAD			
72(48) FCBEOBR FCBNCP	FCBEOBRA			(BASM,BPAM INTERFACE)
76(4C) FCBIOBR	FCBEOBRA			
80(50) FCBDIRDA	81(51) FCBDIRIM			
84(54) FCBCNTRL/FCBNOTE/FCBPOINT	FCBCNTRA/FCBNOTEA/FCBPOINA			
88(58) reserved				
92(5C) reserved				
96(60) FCBOPTEX				

Table 1. SAMPLE OF FCB(File Control Block)

Notes

[1] Cooley, M (1987) *Architect or Bee?* Translated into Japanese by F. Satofuka, I. Inoo & Y. Sato

[2] Kobayasi, I & Oda, T (1986) *Computer History* Ohm

[3] Facom OSIV/X8 Handbook 1980. Fujitsu Co. Ltd.

[4] Takahashi, N (1989) A Programming Environment in the Japanese Language. In *Information Processing* Vol 30 nr 4, pp. 363-372

[5] Fujiysu (1984) *The Criterion of Productivity* Fujitsu Co. Ltd.

[6] Rosenbrock, H H (1988) Engineering As An Art. In *AI & Society* Vol 2(4)